Annals of Mathematics Studies

Number 60

ANNALS OF MATHEMATICS STUDIES

Edited by Robert C. Gunning, John C. Moore, and Marston Morse

1. Algebraic Theory of Numbers, *by* HERMANN WEYL
3. Consistency of the Continuum Hypothesis, *by* KURT GÖDEL
11. Introduction to Nonlinear Mechanics, *by* N. KRYLOFF *and* N. BOGOLIUBOFF
20. Contributions to the Theory of Nonlinear Oscillations, Vol. I, *edited by* S. LEFSCHETZ
21. Functional Operators, Vol. 1, *by* JOHN VON NEUMANN
24. Contributions to the Theory of Games, Vol. I, *edited by* H. W. KUHN *and* A. W. TUCKER
25. Contributions to Fourier Analysis, *edited by* A. ZYGMUND, W. TRANSUE, M. MORSE, A. P. CALDERON, *and* S. BOCHNER
27. Isoperimetric Inequalities in Mathematical Physics, *by* G. PÓLYA *and* G. SZEGÖ
28. Contributions to the Theory of Games, Vol. II, *edited by* H. W. KUHN *and* A. W. TUCKER
30. Contributions to the Theory of Riemann Surfaces, *edited by* L. AHLFORS *et al.*
33. Contributions to the Theory of Partial Differential Equations, *edited by* L. BERS, S. BOCHNER, *and* F. JOHN
34. Automata Studies, *edited by* C. E. SHANNON *and* J. MCCARTHY
38. Linear Inequalities and Related Systems, *edited by* H. W. KUHN *and* A. W. TUCKER
39. Contributions to the Theory of Games, Vol. III, *edited by* M. DRESHER, A. W. TUCKER *and* P. WOLFE
40. Contributions to the Theory of Games, Vol. IV, *edited by* R. DUNCAN LUCE *and* A. W. TUCKER
41. Contributions to the Theory of Nonlinear Oscillations, Vol. IV, *edited by* S. LEFSCHETZ
42. Lectures on Fourier Integrals, *by* S. BOCHNER
43. Ramification Theoretic Methods in Algebraic Geometry, *by* S. ABHYANKAR
44. Stationary Processes and Prediction Theory, *by* H. FURSTENBERG
45. Contributions to the Theory of Nonlinear Oscillations, Vol. V, *edited by* L. CESARI, J. LASALLE, *and* S. LEFSCHETZ
46. Seminar on Transformation Groups, *by* A. BOREL *et al.*
47. Theory of Formal Systems, *by* R. SMULLYAN
48. Lectures on Modular Forms, *by* R. C. GUNNING
49. Composition Methods in Homotopy Groups of Spheres, *by* H. TODA
50. Cohomology Operations, *lectures by* N. E. STEENROD, *written and revised by* D. B. A. EPSTEIN
51. Morse Theory, *by* J. W. MILNOR
52. Advances in Game Theory, *edited by* M. DRESHER, L. SHAPLEY, *and* A. W. TUCKER
53. Flows on Homogeneous Spaces, *by* L. AUSLANDER, L. GREEN, F. HAHN, *et al.*
54. Elementary Differential Topology, *by* J. R. MUNKRES
55. Degrees of Unsolvability, *by* G. E. SACKS
56. Knot Groups, *by* L. P. NEUWIRTH
57. Seminar on the Atiyah-Singer Index Theorem, *by* R. S. PALAIS
58. Continuous Model Theory, *by* C. C. CHANG *and* H. J. KEISLER
59. Lectures on Curves on an Algebraic Surface, *by* DAVID MUMFORD
60. Topology Seminar, Wisconsin, 1965, *edited by* R. H. BING *and* R. J. BEAN

TOPOLOGY SEMINAR WISCONSIN, 1965

S. ARMENTROUT
R. H. BING
C. E. BURGESS
J. L. BRYANT
A. C. CONNOR
J. DANCIS
E. FADELL
J. B. FUGATE
C. H. GIFFEN
H. GLUCK
R. W. HEATH
D. W. HENDERSON
F. B. JONES
L. LININGER
J. MARTIN
L. F. McAULEY
D. R. McMILLAN, JR.
D. V. MEYER
M. E. RUDIN
N. SMYTHE
J. STALLINGS
R. H. SZCZARBA
E. S. THOMAS, JR.
P. A. TULLEY
G. S. UNGAR
J. N. YOUNGLOVE

EDITED BY

R. H. Bing and R. J. Bean

PRINCETON, NEW JERSEY
PRINCETON UNIVERSITY PRESS
1966

L. C. Card 66-21829

Printed in the United States of America
by Westview Press, Boulder, Colorado

Princeton University Press On Demand Edition, 1985

DEDICATION

The contributors to this volume dedicate their work to the memory of

M. K. Fort, Jr.

whose warmth and good will have been felt by the entire mathematical community.

CONTENTS

PREFACE

This volume contains the proceedings of the topology seminar held under the direction of R. H. Bing at the University of Wisconsin in the summer of 1965. Most of the articles were presented as talks to the topologists assembled, then expanded to make expository articles with the details of proofs frequently omitted.

No effort was made to control the subject matter of the talks. On the contrary, each participant was invited to discuss the topic of his current interest. The chapter headings are mine and I confess having to stretch a point here and there to make the articles fit.

An attempt was made to include in the book as many questions and conjectures as possible. Most of these fit nicely into the articles and were included by the authors. Still others appear at the end of the chapters. Most of the questions were asked at special "conjecture sessions" of the seminar.

Following is a list of the graduate staff present at the seminar. The last three names are those of students who contributed to this volume. In addition to those named, the success of the seminar was due, in a large measure, to the students who were present. Their participation, especially in the numerous discussions which followed the talks, was greatly appreciated and, I am sure, mutually beneficial.

Armentrout, Steve		University of Iowa
Bean, Ralph J.	University of Wisconsin	University of Tennessee
Bing, R. H.		University of Wisconsin
Bryant, John	University of Georgia	University of Mississippi
Fadell, Edward		University of Wisconsin
Fugate, Joseph		University of Kentucky
Giffen, Charles		Institute for Advanced Study
Gluck, Herman		Harvard University
Heath, Robert	University of Georgia	Arizona State University
Henderson, David		Institute for Advanced Study
Husseini, Sufien		University of Wisconsin
Jones, Burton	University of California, Riverside, California	

Kister, James	University of Michigan
Lehner, Guydo	University of Maryland
Lininger, Lloyd	University of Missouri
Martin, Joseph	University of Wisconsin
McAuley, Louis	Rutgers University
McMillan, Russell	University of Virginia
Meyer, Donald	Central College
Roy, Prabir	University of Wisconsin
Rudin, Mary Ellen	University of Wisconsin
Smythe, N.	University of New South Wales, Sydney, Australia
Stallings, J.	Princeton University
Szczarba, R.	Yale
Thomas, E. S.	University of Michigan
Tulley, P.	University of Maryland
Younglove, James	University of Houston
Connor, A. C.	University of Georgia
Dancis, Jerome	University of Wisconsin
Ungar, Gerald	Rutgers University

We all express our gratitude to the secretarial staff at both the University of Wisconsin and the University of Tennessee, especially Mrs. Margaret Blumberg and Mrs. Sue Hart.

Ralph Bean

University of Tennessee

CHAPTER I: 3-MANIFOLDS

MONOTONE DECOMPOSITIONS OF E^3

by

Steve Armentrout*

§ 1. Introduction

In 1924, R. L. Moore introduced the notion of an upper semi-continuous collection of point-sets [63] and in 1925, there appeared his proof [64] of the following result:

THEOREM (Moore). If G is an upper semi-continuous decomposition of the Euclidean plane E^2 into compact continua such that no element of G separates E^2, then the decomposition space associated with G is homeomorphic to E^2.

In the following ten years, Moore [65, 66], Alexandroff [1], Kuratowski [50], and others made significant contributions to the theory of upper semi-continuous decompositions. But none of these results dealt specifically with upper semi-continuous decompositions of E^3 into compact continua.

G. T. Whyburn, in an address to the American Mathematical Society in 1935 [80], raised the question of finding conditions on an upper semi-continuous decomposition of E^3 sufficient to insure that the associated decomposition space is homeomorphic to E^3. He pointed out that even if the decomposition has just one non-degenerate element and that element is an arc, some additional condition is necessary. Antoine had constructed [4] an arc α in E^3 such that $E^3 - \alpha$ is not homeomorphic to the complement, in E^3, of a one-point set. If G is the upper semi-continuous decomposition of E^3 whose only non-degenerate element is α, then the decomposition space associated with G is not homeomorphic to E^3.

Whyburn proposed [80, p. 70] imposing the following "trial condition" on each non-degenerate element g of an upper semi-continuous decomposition G of E^3 into compact continua: $E^3 - g$ is homeomorphic to the complement, in E^3, of a one-point set. Further, this is consistent with the Moore theorem on the plane since a compact continuum g in E^2 fails to separate E^2 if and only if $E^2 - g$ is homeomorphic to the complement, in E^2, of a one-point set.

* Research supported by National Science Foundation Grant No. GP-4508.

The "trial condition" proposed by Whyburn came to be known as "point-like" [19] and thus we can state Whyburn's question as follows:

Is it true that if G is an upper semi-continuous decomposition of E^3 into point like compact continua, then the decomposition space associated with G is homeomorphic to E^3 ? An affirmative answer to this question would yield a "natural" generalization to E^3 of the Moore theorem on decompositions of E^2 stated above.

Wardwell, using conditions similar to these, established [78] a condition sufficient for the topological equivalence of a compact metric space X and the space of a monotone decomposition of X. He obtained some theorems concerning point-like decompositions of E^3.

In 1957 Bing gave [20] a negative answer to Whyburn's question. One of the fundamental questions in the study of monotone decompositions of E^3 has accordingly been the following:

For what monotone decompositions of E^3 is the associated decomposition space homeomorphic to E^3 ?

The investigation of this fundamental question has led to the raising of a number of related questions concerning monotone decompositions of E^3, S^3, and 3-manifolds. The purpose of this article is to give a comprehensive survey of known results dealing with these various questions.

In section 3, we consider results dealing with the problem of finding conditions on monotone decompositions of E^3 under which the associated decomposition space is homeomorphic to E^3. In section 4, we consider a condition useful in showing that certain decomposition spaces are homeomorphic to E^3. Section 5 deals with the existence of pseudo-isotopies. In section 6, we consider some conditions which imply that the elements of a decomposition are point-like.

Section 7 deals with the problem of characterizing spaces of monotone decompositions of E^3. Section 8 concerns local properties of decomposition spaces. In section 9, we consider homology and homotopy of decomposition spaces. Recent results by T. M. Price and others on the relationship between the homotopy of a space and that of an associated decomposition space have led to some interesting results. Dimension of decomposition spaces is the subject of section 10. In section 11, we consider the problem of embedding decomposition spaces of euclidean spaces in euclidean spaces.

Products of decomposition spaces with E^1 and with other spaces is the subject of section 12. In section 13, we consider monotone decompositions of E^3 that yield 3-manifolds. Section 14 mentions some closely related problems.

§ 2. Notation and Terminology

Suppose X is a topological space. The statement that G is a _decomposition_ of X means that G is a collection of subsets of X such that each point of X belongs to one and only one set of G.

Suppose X is a topological space. The statement that G is an _upper semi-continuous decomposition_ of X means that G is a decomposition of X such that if g is any element of G and U is any open set in X containing g, then there is an open set V in X such that $g \subset V$, $V \subset U$, and V is a union of elements of G.

Suppose X is a topological space and G is an upper semi-continuous decomposition for X. Let B be the set of all subsets W of G such that the union of all the elements of W is an open subset of X. Then B is a base for a topology T for the set G, and (G, T) is the _decomposition space_ assiciated with G. The decomposition space associated with G will be denoted by X/G.

Throughout this paper, P denotes the projection map from X onto X/G (if $x \in X$, P(x) is the set of G to which x belongs). P is a closed continuous function.

H_G denotes the union of all the non-degenerate elements of G.

For a presentation of a number of fundamental results on upper semi-continuous decompositions of metric spaces into compact continua, see [67, Chap. V] or [81, Chap. 7]. We shall merely state a few of the most useful results concerning such decompositions. Suppose then that X is a _separable_ metric space and G is an upper semi-continuous decomposition of X into compact continua. Then the following hold:

1. X/G is a separable metric space.
2. If X is locally compact, so is X/G.
3. If X is locally connected, so is X/G.
4. If M is a compact subset of X/G, then $P^{-1}[M]$ is compact.
5. If each set of G is connected and M is a connected subset of X/G, then $P^{-1}[M]$ is connected.

There is a frequently used process for identifying a decomposition space. Suppose X is a topological space and G is an upper semi-continuous decomposition of X. Suppose further that f is a continuous function from X onto a space Y such that $G = \{f^{-1}[y]: y \in Y\}$. There is a natural map from X/G onto Y, namely fP^{-1}, and we shall let φ denote this map. It would be convenient in certain cases if φ is a homeomorphism from X/G onto Y. For information on this question, see [81, Chap.7], [82], and [44, Chap. 3].

Suppose now that X is a metric space. The statement that G is a _monotone_ decomposition of X means that G is an upper semi-continuous decomposition of X into compact continua. In this article, we are concerned only with monotone decompositions.

Suppose now that n is some positive integer and X is either E^n or S^n. The statement that the compact continuum A of X is point-like (in X) means that $X - A$ is homeomorphic to the complement, in X, of a one-point set.

Suppose that M is an n-manifold. The subset A of M is cellular (in M) if and only if there exists a sequence $C_1, C_2, C_3, \ldots$ of n-cell in M such that for each positive integer i, $C_{i+1} \subset \text{Int } C_i$ and $A = \cap_{i=1}^{\infty} C_i$.

It is known that if X is either E^n or S^n, "point-like" and "cellular" are equivalent for compact continua in X. For a proof of this, see [76]. A different proof, not dependent on [22], can be obtained by following [77].

Suppose that X is E^n or S^n. The statement that G is a point-like decomposition of X means that G is an upper semi-continuous decomposition of X such that each element of G is a point-like compact continuum in X. If M is an n-manifold, G is a cellular decomposition of M if and only if G is an upper semi-continuous decomposition of M such that each element of G is cellular.

§3. Decompositions of E^3 that yield E^3

In 1957, Bing gave an example [20] of a decomposition of E^3 into tame arcs and one-point sets such that the associated decomposition space is not homeomorphic to E^3. Since the appearance of [20], there have been numerous results dealing with the question of when a monotone decomposition of E^3 has a decomposition space homeomorphic to E^3.

a. The countable case

First we shall consider the case where the decomposition has only countably many non-degenerate elements. We start by stating some conditions sufficient in this case.

Bing proved [19] that if G is a monotone decomposition of E^3 having only countably many non-degenerate elements, then E^3/G is homeomorphic to E^3 provided G satisfies any one of the following conditions:

(1) Each element of G is point-like and H_G is a G_δ.

(2) Each element of G is starlike, i.e., if $g \in G$, there is a point p of g such that if ℓ is a line through p, $\ell \cap g$ is an interval or a one-point set.

(3) Each non-degenerate element of G is a tame arc.

Each tame arc is point-like and each starlike compact continuum is point-like.

A collection C of subsets of a metric space is a <u>null collection</u> if and only if for each positive number ε, there exist, at most, finitely many sets of C of diameter more than ε.

Meyer proved [61] that if G is a monotone decomposition of E^3 into a null collection of tame 3-cells and one-point sets, then E^3/G is homeomorphic to E^3. Bean announced [16] that a monotone decomposition into a null family of star-like equivalent sets yields E^3 as its decomposition space; a star-like equivalent set is one homeomorphic to a star-like set, under a space homeomorphism.

Gillman and Martin announced [41] that if G is a monotone decomposition of E^3 into one-point sets and countably many arcs, each locally wild at only finitely many points, then E^3/G is homeomorphic to E^3.

Now we shall consider conditions not sufficient to insure that the decomposition space will be homeomorphic to E^3. Bing gave [25] an example of a decomposition G into countably many planar point-like sets and one-point sets such that E^3/G is not homeomorphic to E^3. In fact, there exist two planes such that each non-degenerate element of G lies in one of these planes. This example shows that in condition (1) above, the condition that H_G be a G_δ is a necessary one.

In [25], Bing raised the following question: If G is a point-like decomposition of E^3 with only countably many non-degenerate elements and each non-degenerate element is an arc, is E^3/G homeomorphic to E^3 ? Gillman and Martin announced [41] a negative solution to this question.

It was announced [6] that there is a monotone decomposition G of E^3 into tame 3-cells and one-point sets such that E^3/G is not homeomorphic to E^3. By modifying G, one could obtain monotone decompositions G_1, of E^3 into countably many tame discs and one-point sets, and G_2 into countably many tame dendrons and one-point sets such that neither E^3/G_1, nor E^3/G_2 is homeomorphic to E^3. However, an error was found in the argument that E^3/G is not homeomorphic to E^3, so the following questions are open at the present:

<u>Question</u> 1. Does there exist a monotone decomposition G of E^3 into tame 3-cells and one-point sets such that E^3/G is not homeomorphic to E^3 ?

<u>Question</u> 2. Does there exist a monotone decomposition G of E^3 into, at most, countably many discs and one-point sets such that E^3/G is not homeomorphic to E^3 ?

<u>Question</u> 3. Does there exist a monotone decomposition G of E^3 into one-point sets and at most, countably many tame dendrons such that E^3/G is not homeomorphic to E^3 ? Into triods? n-ods ?

There are some conditions known to be both necessary and sufficient in the case we are considering in order that the decomposition space be homeomorphic to E^3.

Kwun [51] gives conditions in terms of I-sets. A subset F of the 3-sphere S^3 is an I-set if and only if there is a countable set A in S^3 such that $S^3 - F$ is homeomorphic to $S^3 - A$. Kwun proves that if G is a monotone decomposition of S^3 having only countably many non-degenerate elements, then S^3/G is a 3-sphere if and only if H_G is an I-set.

Price proved [68] that if G is a monotone decomposition of E^3 with only countably many non-degenerate elements, then E^3/G is homeomorphic to E^3 if and only if for each non-degenerate element g of G and each open set V containing g, there is a 3-cell C in E^3 such that $g \subset \text{Int } C$, $C \subset V$, and Bd C and H_G are disjoint. The sufficiency of this condition holds in all cases; see part c of this section. Using the methods of [68] or [5], one may show that the 3-cell C can be required to be tame.

A necessary and sufficient condition for the case considered and closely related to that just above may be found in [5]. Additional necessary and sufficient conditions can be found in sections 4 and 5.

Some of the results of [13] make the countable case of more interest than just one where there is a powerful and useful hypothesis. It is proved in [13] that there is a point-like decomposition F of E^3 into one-point sets and, at most, countably many compact continua such that F and the dogbone decomposition G [20] are equivalent (in the sense of [13]). Since E^3/F and E^3/G are homeomorphic, it may be possible to use some of the results of this section to give another proof that the dogbone decomposition space is not homeomorphic to E^3.

b. The compact 0-dimensional case

Now we consider monotone decompositions G of E^3 such that $P[H_G]$ is contained in a compact 0-dimensional set.

A major reason for interest in this case is that many examples satisfy the hypothesis for this case. If one constructs a monotone decomposition of E^3 in a fairly "natural" way, then the resulting example is of the type under consideration. This method of construction is as follows: Let M_1, M_2, M_3,... be a sequence of compact 3-manifolds-with-boundary in E^3 such that if i is any positive integer, $M_{i+1} \subset \text{Int } M_i$. Now define G to be the decomposition of E^3 such that g is a non-degenerate element of G if and only if g is a non-degenerate component of $\cap_{i+1}^{\infty} M_i$. G is upper semi-continuous [67, Chap. V] and it is not hard to see that $P[H_G]$ is contained in a compact 0-dimensional set, namely $P[\cap_{i=1}^{\infty} M_i]$. $P[H_G]$ itself may be countable, as the example of section 3 of [25] shows.

Since sequences such as M_1, M_2, M_3,... are frequently useful, we introduce a name for such sequences. Suppose G is a monotone decomposition of E^3 such that $P[H_G]$ is contained in a compact 0-dimensional set. Then M_1, M_2, M_3,... is a defining sequence for G if and only if M_1, M_2,

$M_3, \ldots$ is a sequence of compact 3-manifolds-with-boundary in E^3 such that:

(1) For each positive integer i, $M_{i+1} \subset \operatorname{Int} M_i$, and

(2) g is a non-degenerate element of G if and only if g is a non-degenerate component of $\cap_{i=1}^{\infty} M_i$.

It can be proved that if G is a monotone decomposition of E^3 such that $P[H_G]$ is contained in a compact 0-dimensional set, then there is a defining sequence for G. Hence it follows that if G is a monotone decomposition of E^3, then $P[H_G]$ is contained in a compact 0-dimensional set if and only if there is a defining sequence for G.

If G is a monotone decomposition of E^3 such that $P[H_G]$ is a Cantor set, it is sometimes said that G has a Cantor set of non-degenerate elements.

Among the examples of monotone decompositions of E^3 that have appeared in the literature and satisfy the hypothesis of this part of section 3 are [20], [25], and [12]. The proposed examples [6] and [58] also satisfy this hypothesis.

Bing proved [17] that the sum of two Alexander horned spheres, sewed together on their boundaries by the identity, is a 3-sphere by considering a decomposition of E^3 with a Cantor set of arcs as non-degenerate elements. Casler [32] uses similar ideas to show that for any sewing of the boundaries of two Alexander horned spheres, the result is a 3-sphere.

In fact, the decomposition used by Bing in [17] is a toroidal decomposition. G is a toroidal decomposition of E^3 if and only if G is a monotone decomposition of E^3 such that there is a defining sequence M_1, M_2, $M_3, \ldots$ for H_G such that for each positive integer i, each component of M_1 is a solid torus (3-cell with one handle). Keldy's announced [46] that if G is a toroidal decomposition of E^3 into point-like locally connected continua such that the set of all non-degenerate elements is a continuous collection, then E^3/G is homeomorphic to E^3. Bing and Armentrout showed [12] that there is a toroidal decomposition G of E^3 into tame arcs and one-point sets such that E^3/G is not homeomorphic to E^3.

The example of section 3 of [25] is a toriodal decomposition, and a modification of this example described in [13] yields a toroidal decomposition G of E^3 into point-like continua and one-point sets such that E^3/G is not homeomorphic to E^3.

In [20], Bing described a decomposition G into tame arcs and one-point sets such that E^3/G is not homeomorphic to E^3 (the dogbone decomposition). Fort showed [39] that there exists such a decomposition in which each arc is polygonal. An interesting decomposition equivalent to the dogbone decomposition may be seen in [13].

The following question is still apparently open:

Question 4. Is it true that if G is a monotone decomposition of E^3 into straight-line intervals and one-point sets, then E^3/G is homeomorphic to E^3 ?

This question was raised by Bing and in section 6 of [25] and section 10 of [27], he describes an example which may provide a negative solution to Question 4. However, it is not known now whether the decomposition space of Bing's example is topologically different from E^3. McAuley announced [58] an example to show that Question 4 has a negative answer, but for it also it is not known whether the decomposition space is topologically distinct from E^3. Both Bing's example and McAuley's example have Cantor sets of non-degenerate elements.

Since there are wild Cantor sets in E^3, the following problem seems interesting: Suppose C is a compact 0-dimensional set in E^3. Is there a point-like decomposition G of E^3 such that E^3/G is homeomorphic to E^3 and a homeomorphism h from E^3/G onto E^3 such that h sends $P[H_G]$ onto C ? In other words, not only is $P[H_G]$ homeomorphic to C, but it is embedded in the decomposition space just like C is embedded in E^3. In [8], it is shown that for any such C, there is such a decomposition G.

c. General case

Dyer and Hamstrom proved [37] that if G is a point-like decomposition of E^3 such that each element of G lies in a horizontal plane, then E^3/G is homeomorphic to E^3.

In connection with Question 4, McAuley proved [56] that if G is a monotone decomposition of E^3 into straight-line intervals and one-point sets such that there exist countably many discs D_1, D_2, $D_3, \ldots$ such that each interval of G intersects and is perpendicular to some one of D_1, D_2, $D_3, \ldots$, then E^3/G is homeomorphic to E^3.

McAuley has given in [57] some general conditions under which monotone decompositions of E^3 yield E^3 as decomposition spaces.

There are some additional results for arbitrary monotone decompositions similar to some of those stated in part (a) for the countable case. In [56], McAuley gave a condition sufficient to show that if G is a point-like decomposition of E^3 such that H_G is a G_δ, then E^3/G is homeomorphic to E^3. McAuley's condition involves coverings of H_G by certain kinds of collections of tame 3-cells. Price proved [68] that if G is a monotone decomposition of E^3 and for each non-degenerate element g of G and each neighborhood V of G, there is a 3-cell C in E^3 such that $g \subset \operatorname{Int} C$, $C \subset V$, and $\operatorname{Bd} C$ and H_G are disjoint, then E^3/G is homeomorphic to E^3. A theorem proved by Smythe [75] also provides sufficient condition for the space of a monotone decomposition of E^3 to be homeomorphic to E^3.

d. Remarks

An interesting example due to Bing [27, section 4] leads one to suspect that for arbitrary monotone decompositions of E^3 and S^3, the situation is complicated. Bing describes a monotone decomposition G of S^3 such that S^3/G is a 3-sphere but no non-degenerate element of G is point-like. In this case, $P[H_G]$ is an arc.

Most of the results of this section deal with monotone decompositions G of E^3 (or S^3) such that $P[H_G]$ is a 0-dimensional set. Difficulties of a type not yet studied appear in case $P[H_G]$ has non-degenerate components.

It would be of interest to have an answer to the following:

Question 5. Suppose G is a monotone decomposition of E^3 into arcs and one-point sets. Suppose E^3/G is homeomorphic to E^3 and $P[H_G]$ is a 2-sphere. Under what conditions is H_G a spherical shell (i.e., homeomorphic to the Cartesian product of a 2-sphere and $[0, 1]$). If it helps, assume $P[H_G]$ is a tame 2-sphere, and that each arc of G is tame. Is it sufficient to assume that the set of all non-degenerate elements of G is a continuous collection ?

Roberts proved [69] that there is no monotone decomposition of E^2 into arcs.

Question 6. Is there a monotone decomposition of E^3 into arcs?

If a monotone decomposition G of E^3 has the property that $P[H_G]$ is either countable or lies in a compact 0-dimensional set, one may be able to select certain 3-manifolds-with-boundary in E^3 whose interiors contain non-degenerate elements of G, whose boundaries miss H_G, and whose images under P are small. If $P[H_G]$ has non-degenerate components, this process fails, at least for some non-degenerate elements. It may, therefore, be fruitful to study monotone decompositions G of compact 3-manifolds-with-boundary M such that some non-degenerate elements of G intersect Bd M.

Question 7. Suppose C is a 3-cell and G is a monotone decomposition of C such that each element of G intersects Bd C. When is C/G a 3-cell? For an informative example where C/G is not a 3-cell, see [18]. Methods related to those of [18] may be useful here.

The notion of "lifting" a set under P^{-1} has proved useful in the study of decompositions of E^3 that yeild 3-manifolds as decomposition space. The following question suggests a similar technique.

Question 8. Suppose G is a point-like decomposition of E^3. If S is a 2-sphere in E^3/G, does there exist a 2-sphere S' in E^3 such that $P[S']$ is a 2-sphere homeomorphically close to S ?

§4. A useful condition

In this section we consider a condition that is useful in the study of decomposition spaces. It was introduced by Bing in [17] and used in [17] and [19] to show that for certain monotone decompositions of E^3, the associated decomposition space is homeomorphic to E^3. It was used to show that the dogbone decomposition space [20] is not homeomorphic to E^3. Since the appearance of these papers, it has found many applications. In certain cases, a decomposition has a decomposition space homeomorphic to E^3 if and only if this condition holds.

Suppose that G is a monotone decomposition of E^3 such that $P[H_G]$ is 0-dimensional. Then G satisfies <u>condition</u> B if and only if for each open set U in E^3 containing H_G and each positive number ε, there is a homeomorphism h from E^3 onto E^3 such that:

(1) if $x \in E^3 - U$, $h(x) = x$ and

(2) if $g \in G$, (diam $h[g]$) $< \varepsilon$.

There is an apparently stronger condition that is of importance. Again, suppose G is a monotone decomposition of E^3 such that $P[H_G]$ is 0-dimensional. Then G satisfies <u>condition</u> B^* if and only if for each open set U in E^3 containing H_G, each positive number ε, and each homeomorphism f from E^3 onto E^3, there is a homeomorphism h from E^3 onto E^3 such that:

(1) if $x \in E^3 - U$, $h(x) = f(x)$ and

(2) if $g \in G$, then (diam $h[g]$) $< \varepsilon$.

It is a corollary of a result of [10] that if G is a monotone decomposition of E^3 such that $P[H_G]$ is 0-dimensional, then condition B and condition B^* are equivalent.

The following theorem is due to Bing [28].

THEOREM. If G is a monotone decomposition of E^3 such that $P[H_G]$ is 0-dimensional and G satisfies condition B, then E^3/G is homeomorphic to E^3.

A proof of this theorem may be obtained by following the proof of Theorem 1 of [19] and using the fact that, under the hypothesis, condition B^* holds.

Now we shall consider two cases in which one can show that if the decomposition space is homeomorphic to E^3, then condition B holds. This line of argument is used in [20] and [12] to show that certain decomposition spaces are not homeomorphic to E^3.

Suppose G is a monotone decomposition of E^3 such that G has only countably many non-degenerate elements. It was proved in [5] that if E^3/G is homeomorphic to E^3, then G satisfies condition B. The proof of this is based on the following result:

THEOREM. Suppose G is a monotone decomposition of E^3 such that $P[H_G]$ is countable and E^3/G is homeomorphic to E^3. If S is a tame 2-sphere in E^3/G such that S and $P[H_G]$ are disjoint, then $P^{-1}[S]$ is tame in E^3.

A proof of this result, simpler than that of [5], can be constructed following [68].

A similar result holds in the compact 0-dimensional point-like case. In [8], it is shown that if G is a point-like decomposition of E^3 such that $P[H_G]$ is contained in a compact 0-dimensional set and E^3/G is homeomorphic to E^3, then condition B holds. The proof is based on the following result established in [8]:

THEOREM. Suppose G is a point-like decomposition of E^3 such that $P[H_G]$ is contained in a compact 0-dimensional set and E^3/G is homeomorphic to E^3. Suppose M is a compact 3-manifold-with-boundary in E^3 such that Bd M and $C\ell\ H_G$ are disjoint. Then there is a homeomorphism h from M onto $P[M]$ such that $h|$ Bd $M = P|$ Bd M.

Condition B is a useful condition in the study of monotone decompositions G of E^3 such that $P[H_G]$ is 0-dimensional. It appears not to be particularly useful if $P[H_G]$ has a non-degenerate component. One may seek an analogous condition for other cases. It seems likely that a condition involving open coverings of $P[H_G]$ would prove to be useful in some cases. Some conditions of this nature have been used by McAuley [56].

<u>Question</u> 9. Suppose G is a point-like decomposition of E^3 such that $P[H_G]$ is 0-dimensional and E^3/G is homeomorphic to E^3. Does G satisfy condition B ?

§ 5. Existence of pseudo-isotopies

If H is a homotopy from $E^3 \times I$ into E^3 and $t \in I$, then H_t denotes the function from E^3 into E^3 such that if $x \in E^3$, $H_t(x) = H(x, t)$. The statement that φ is an <u>isotopy</u> from $E^3 \times I$ into E^3 means that φ is a homotopy from $E^3 \times I$ into E^3 such that if $t \in I$, φ_t is a homeomorphism. The statement that φ is a <u>pseudo-isotopy</u> from $E^3 \times I$ into E^3 such that if $t \in [0, 1)$, φ_t is a homeomorphism.

It is of interest to know whether, for a given monotone decomposition G of E^3 such that E^3/G is homeomorphic to E^3, there exists a pseudo-isotopy φ from $E^3 \times I$ into E^3 and a homeomorphism h from E^3/G onto E^3 such that if $g \in G$, $h(g) = \varphi_1[g]$.

The example of Bing's described in section 4 of [27] is an example of such a monotone decomposition of E^3 for which no such pseudo-isotopy exists.

Bing pointed out in [19] that his argument for Theorem 1 of [19] may, in certain cases, be modified to show the existence of pseudo-isotopies of the type described above. With the aid of such a modification, it was shown [5] that if G is a monotone decomposition of E^3 with only countably many non-degenerate elements and such that E^3/G is homeomorphic to E^3, there is a pseudo-isotopy of the type specified above.

Bean has shown [15] that if G is a monotone decomposition of E^3 such that:

(1) H_G is definable by cells-with-handles (see sec 6), and

(2) E^3/G is homeomorphic to E^3, then there is a pseudo-isotopy as described above.

§6. The cellularity problem

The following question arises naturally in the study of monotone decompositions of E^3: Is it true that if G is a monotone decomposition of E^3 such that E^3/G is homeomorphic to E^3, then each element of G is cellular?

The example of Bing's [27, section 4] mentioned previously gives a negative answer to this question. Thus we are lead to restrict G and ask the following:

Question 10. Suppose G is a monotone decomposition of E^3 such that $P[H_G]$ is 0-dimensional and E^3/G is homeomorphic to E^3. Is each element of G cellular?

There are some partial affirmative solutions for Question 10. It was proved by Kwun [51] that if G is a monotone decomposition of E^3 such that E^3/G is homeomorphic to E^3 and G has only countably many non-degenerate elements, then each element of G is cellular.

Now suppose that G is a monotone decomposition of E^3 such that E^3/G is homeomorphic to E^3 and $P[H_G]$ is contained in a compact 0-dimensional set. As noted in part b) of section 3, this implies that G has a defining sequence $M_1, M_2, M_3, \ldots$. If $P[H_G]$ is contained in a tame Cantor set (see [26] for definition) in E^3/G, then it is easy to see H_G has a defining sequence $M_1, M_2, M_3, \ldots$ such that for each positive integer i, each component of M_i is a 3-cell. In this case, each element of G is cellular.

We shall say that H_G is definable by 3-cells-with-handles if and only if there is a defining sequence $M_1, M_2, M_3, \ldots$ for H_G such that for each positive integer i, each component of M_i is a 3-cell-with-handles. Bean has proved [15] that if H_G is definable by 3-cells-with-handles, then each element of G is cellular.

If M is a compact 3-manifold-with-boundary, then one may construct a 3-cell-with-handles from M in the following way: Suppose T is a triangulation of M and let T_1 be the carrier of the 1-skeleton of T. By constructing a suitable tubular neighborhood L of T_1, one may show that $C\ell(M - L)$ is a 3-cell-with-handles. To show that, for a given monotone decomposition G of E^3 such that $P[H_G]$ lies in a compact 0-dimensional set, H_G is definable by 3-cells-with-handles, one considers a defining sequence $M_1, M_2, M_3, \ldots$ for H_G. If for each i and each component M_{ij} of M_i, there exists a triangulation T_{ij} of M_{ij} such that the carrier of the 1-skeleton of T_{ij} is disjoint from $C\ell\ H_G$, then $(C\ell H_G) \cap M_{ij}$ lies in a 3-cell-with-handles K_{ij} contained in M_{ij}.

The following is a condition under which there exist triangulations of the desired type: If g is an element of G, g_0 is any subcontinuum of g embeddable in E^2, and h is some embedding of g_0 in E^2, then $h[g_0]$ does not separate E^2. Since separating E^2 is a topological property of a compact continuum, this condition is meaningful. As a corollary of this result, it follows that, under the hypotheses on G stated above, each element of G is cellular provided either

(1) each element of G is tree-chainable,

(2) each non-degenerate element of G is a dendron, or, in particular,

(3) each non-degenerate element of G is an arc.

It follows from Corollary 2 of [53] that if G is a decomposition of E^3 into one-point sets and simple closed curves such that $P[H_G]$ is a compact 0-dimensional set, then E^3/G is not homeomorphic to E^3. The results of [53] may be useful in attacking Questions 10 and 11. In some cases, a monotone decomposition is equivalent (in the sense of [13]) to a decomposition into 1-dimensional continua: see [13].

<u>Question</u> 11. Suppose G is a monotone decomposition of E^3 such that E^3/G is homeomorphic to E^3, $P[H_G]$ is contained in a compact 0-dimensional set, and each element of G is 1-dimensional. Is each element of G cellular?

Martin recently proved [55] that if G is a monotone decomposition of E^3 into compact absolute retracts such that E^3/G is a 3-manifold, then each element of G is cellular. His proof uses a criterion for cellularity established by McMillan [59].

It was proved in [8] that if G is a monotone decomposition of E^3 such that $P[H_G]$ lies in a compact 0-dimensional set and G satisfies condition B (see section 4 for definition), then each element of G is cellular.

§7. Characterizing decomposition spaces

R. L. Moore characterized the spaces of monotone decompositions of S^2 [65]. He proved that each monotone decomposition of S^2 has as its decomposition space a cactoid (see [65] for definition), and conversely, given any cactoid, there exists a monotone decomposition G of S^2 such that S^2/G is homeomorphic to K. For the plane, analogous results hold.

A theorem proved by Hurewicz makes it seem unlikely that there is any concise topological characterization of spaces of monotone decompositions of E^3. Hurewicz showed [45] that if X is any compact metric space, there is a monotone decomposition G of E^3 such that E^3/G has a subspace homeomorphic to X.

Even in the case of point-like decompositions of E^3, it appears possible to obtain a variety of distinct decomposition spaces. Little has been done in the direction of showing that various non-euclidean decomposition spaces are topologically distinct. It is a result of [13] that the dogbone decomposition space and the space of the decomposition of section 3 of [25] are not homeomorphic. The methods of [13] may be useful in distinguishing between other pairs of decomposition spaces.

It follows from a result of Wilder's [84] that the space of a point-like decomposition of S^3 is a compact generalized 3-manifold [83]. This result appears to be the best result at the present on the problem of characterizing spaces of point-like decompositions.

It is easy to find spaces which are not spaces of any monotone decomposition of S^3. It is a corollary of a result of [68] that if G is a monotone decomposition of S^3 such that S^3/G is a 3-manifold, then S^3/G is simply connected. Hence if M is a 3-manifold not simply connected, M is not the space of any monotone decomposition of S^3.

§ 8. Local properties of decomposition spaces

In section 3, we stated a number of results concerning conditions on monotone decompositions of E^3, some of which are sufficient and some of which are not sufficient, to insure that the associated decomposition space should be homeomorphic to E^3. The conditions considered in section 3 were conditions on the non-degenerate elements of G. In this section we consider local topological properties of decomposition spaces.

There are relatively few results concerning local properties of decomposition spaces and we shall consider only two such properties. These two and three others are discussed in section 7 of [27].

(a) Each point of the space has arbitrarily small neighborhoods with 2-sphere boundaries.

Bing proved that the example of section 4 of [27] fails to have this property, and hence the decomposition space is not homeomorphic to E^3. Cannon announced [31] that the dogbone space fails to have this property but later withdrew his announcement.

A theorem proved by Harrold [43] shows that a condition closely related to (a) is sufficient to insure that the space of a monotone decomposition of S^3 be a 3-sphere. Harrold proved that if G is a monotone decomposition of S^3 such that each point of $Cl\ P[H_G]$ has arbitrarily small neighborhoods in S^3/G with boundaries that are 2-spheres disjoint from $Cl\ P[H_G]$, then S^3/G is a 3-sphere.

Question 12. Suppose G is a decomposition of E^3 into one-point sets and a Cantor set of tame arcs. If E^3/G has property (a), is E^3/G homeomorphic to E^3 ?

Question 13. Does there exist a point-like decomposition G of E^3 with only countably many non-degenerate elements such that E^3/G is not homeomorphic to E^3 and E^3/G has property (a) ? Is there such an example such that each point of E^3/G has arbitrarily small neighborhoods in E^3/G with 2-sphere boundaries that miss $P[H_G]$?

(b) Each point of the space has arbitrarily small simply connected open neighborhoods.

M. L. Curtis conjectured that the dogbone decomposition space fails to have property (b). It was announced [7] that Curtis' conjecture is correct. The proof is complicated.

Question 14. If X is the decomposition space of the example of section 3 of [25], does X fail to have property (b) ?

§ 9. Homotopy and homology of decomposition spaces

One of the most useful results concerning decomposition spaces has been the following lemma proved by Price [68]:

LEMMA 1. Suppose G is a point-like decomposition of E^3 and U is a simply connected open set in E^3/G. Then $P^{-1}[U]$ is simply connected.

This lemma has been extended in various ways. It holds for cellular decompositions of 3-manifolds [9]. An analogous lemma for the vanishing of the second homotopy group is established in [9].

It is also proved in [68] that if G is a point-like decomposition of E^3, U is an open set in E^3/G, and U is an open 3-cell, then $P^{-1}[U]$ is an open 3-cell. An immediate corollary is the following theorem [68]: If G is a point-like decomposition of E^3 such that E^3/G is a 3-manifold, then if $g \in G$ and U is an open set in E^3 containing g, there is an open 3-cell V in E^3 such that $g \subset V$, $V \subset U$, and V is a union of sets of G.

Another useful result established by Price in [68] is the following:

LEMMA 2. Suppose G is a monotone decomposition of E^3 such that E^3/G is a 3-manifold. If V is an open set in E^3/G and U is a simply connected subset of E^3 such that $V \subset P[U]$, then each loop in V can be shrunk in $P[U]$.

It follows that if G is a monotone decomposition of a simply connected 3-manifold M such that M/G is a 3-manifold, then M/G is simply connected.

One may ask, in fact, whether every monotone decomposition of E^3 has a decomposition space which is simply connected. Bing announced [21] (also, see section 3 of [27]) that if G is the decomposition of E^3 whose only non-degenerate element is a solenoid, then E^3/G is not simply connected.

Price also proved [68] that if G is a cellular decomposition of E^3 such that E^3/G is a 3-manifold, then E^3/G is contractible.

It follows from a theorem of Smale's [74] that if G is a cellular decomposition of a 3-manifold M such that each element of G is a compact absolute retract and M/G is a 3-manifold K, then if $i = 1$, or 2, $\pi_i(K)$ is isomorphic with $\pi_i(M)$. This was pointed out by Martin [55].

Now we shall consider the homology of decomposition spaces. The class of n-gcm's (compact generalized manifolds; see [83] for definition) has a pleasant property relative to certain types of decompositions. Following Wilder [84], we define a mapping f from a space A into a space B to be <u>n-monotone</u> if and only if for each point b of B and each $r \leqq n$, $H^r(f^{-1}[b]) = 0$. In [84], Wilder proved the following: Suppose S is an orientable n-gcm, S' is an, at most, n-dimensional non-degenerate Hausdorff space, and f is an (n-1)-monotone continuous mapping from S onto S'. Then S' is an orientable n-gcm of the same homology type as S.

Consequently, under (n-1)-monotone maps, the n-gcm's form a closed family (modulo the dimension requirement). This result is reminiscent of a result [79] of Whyburn's: The monotone image of a cactoid is a cactoid.

As Price pointed out in [68], if G is a point-like decomposition of E^3, then the associated projection map P is an acyclic map, i.e., for each n and each point x of E^3/G, $P^{-1}[x]$ has trivial Cech homology groups. As we mentioned in section 7, the space of a point-like decomposition of S^3 is a 3-gcm.

In [35], Curtis and Wilder construct examples of 3-gcm's in S^4 such that the 3-gcm's are not 3-manifolds but have the homotopy type of S^3. These are obtained by shrinking various Fox-Artin arcs [14] to points in S^3 and showing that the resulting decomposition spaces can be embedded in S^4.

It is also shown in [35] that the dogbone decomposition space B is a 3-dimensional homotopy manifold in the sense of Griffith (see [35] for definition). It is also proved that B is LC^∞ and that for each k, $\pi_k(B) = 0$.

Kwun and Raymond have obtained [52] some interesting results concerning the cohomology of decomposition spaces.

§10. Dimension of decomposition spaces

The results of Moore [65] characterizing spaces of monotone decompositions of S^2 and E^2 also show that the dimension of such decomposition spaces is, at most, 2. The result of Hurewicz [45] mentioned previously, shows that there is no analogous result for E^3; the dimension of spaces of monotone decompositions of E^3 may be any positive integer or it may be infinite.

As we mentioned in section 9, if G is a point-like decomposition of E^3, the projection map P is acyclic. Dyer proved [36] that if X and Y are compact metric spaces and f is an acyclic continuous map from X onto Y and Y has finite dimension, then $\dim Y \leqq \dim X$. Hence we have the following:

THEOREM. If G is a point-like decomposition of E^3 and E^3/G has finite dimension, then $(\dim E^3/G) \leqq 3$.

<u>Question</u> 15. Is there a geometric proof of the theorem above? Is there one which shows, in addition, that E^3/G is finite-dimensional?

§11. Embedding decomposition spaces in Euclidean spaces

The characterization of spaces of monotone decompositions of S^2 and E^2 by Moore [65] leads to a solution of the problem of embedding such decomposition spaces in Euclidean space. It is not hard to show that each cactoid can be embedded in E^3. Hence, if G is a monotone decomposition of S^2, S^2/G can be embedded in E^3. Similarly, the space of any monotone decomposition of E^2 can be embedded in E^3.

The result [45] of Hurewicz shows that there exists a monotone decomposition G of E^3 such that $(\dim E^3/G)$ is not finite and hence E^3/G is not embeddable in any Euclidean space. Such a decomposition G has uncountably many non-degenerate elements.

Bing and Curtis gave an example [30] of a monotone decomposition G of E^3 such that G has exactly nine non-degenerate elements and each is a circle, and E^3/G is not embeddable in E^4. Rosen showed [71] that there is such an example with just six circles. Goblirsch showed [42] that for certain monotone decompositions G of E^3 such that G has exactly three non-degenerate elements and each is a circle, E^3/G is embeddable in E^4. If a monotone decomposition G of E^3 has, at most, two non-degenrate elements, E^3/G is embeddable in E^4, according to a result of [30]. Keldyš has announced [49] that if G is any monotone decomposition of E^3 with only finitely many non-degenerate elements, then E^3/G can be embedded in E^5.

Question 16. Suppose G is a monotone decomposition of E^3 such that G has only countably many non-degenerate elements. Is E^3/G embeddable in E^5 ? If H_G is a null family, is E^3/G embeddable in E^5 ?

Question 17. If G is a point-like decomposition of E^3, is E^3/G embeddable in E^4 ?

There are some partial affirmative solutions for Question 16. In [34], Curtis proved an embedding theorem for monotone decompositions of E^3 with simply connected elements, and it follows from this result that the dogbone decomposition space can be embedded in E^4. Keldyš proved [46, 47, 48] that if G is a point-like decomposition of E^3 such that $P[H_G]$ is a compact 0-dimensional set, then E^3/G is embeddable in E^4. It follows from results of [10] that if G is any point-like decomposition of E^3 such that $P[H_G]$ is 0-dimensional, then E^3/G is embeddable in E^4. The embedding is realized as the restriction to $E^3 \times \{0\}$ of the final stage of a pseudo-isotopy from $E^4 \times I$ into E^4 that starts at the identity. As a corollary, it follows that if G is a point-like decomposition of E^3 with only countably many non-degenerate elements, then E^3/G can be embedded in E^4.

§12. Products of decomposition spaces

Bing proved [23] that if B is the dogbone decomposition space, then $B \times E^1$ is homeomorphic to E^4. Thus, E^4 has non-manifold factors.

In order to avoid an ambiguity in terminology, we introduce some new notation. In a number of papers on decompositions, the notation "X/A" has denoted the space obtained from X by collapsing the subset A of X to a point. Since this conflicts with a notation of ours, we introduce the following: If X is a metric space and A is a non-degenerate compact subset of X, let G_A be the upper semi-continuous decomposition of X whose only non-degenerate element is A. Then X/G_A denotes the space obtained by collapsing A to a point.

Using techniques analogous to those of [23], Andrews and Curtis proved [2] that if α is any arc in E^n, then $(E^n/G_\alpha) \times E^1$ is homeomorphic to E^{n+1}. Gillman and Martin announced [41] the following extension of the result of [2]: If G is a monotone decomposition of E^n with only countably many non-degenerate elements, each of which is an arc, then $(E^n/G) \times E^1$ is homeomorphic to E^{n+1}.

Rosen showed [72] that there exists space B' such that B' is not Euclidean at any point but $B' \times E^1$ is homeomorphic to E^4. B' is obtained by a modification of the dogbone decomposition.

Rubin showed [73] that if G is the decomposition of E^3 described in section 3 of [25], $(E^3/G) \times E^1$ is homeomorphic to E^4. Andrews and Rubin [3] have generalized Rubin's result [73] in a way that we now indicate.

A sequence $A_1, A_2, A_3, \ldots$ of locally finite disjoint collections of subsets of E^3 is trivial if for each i,

(1) $\cup\{x: x \in A_{i+1}\} \subset \text{Int}\,[\cup\{x: x \in A_i\}]$,
(2) each set of A_i is a 3-cell-with-handles semi-linearly embedded in E^3, and
(3) if $x \in A_i$, $y \in A_{i+1}$, and $y \subset x$, then the inclusion map j from y into $\text{Int}\,x$ is null-homotopic.

If $A_1, A_2, A_3, \ldots$ is a trivial sequence, let G be the one-point subsets of $E^3 - \cap_{i=1}^{\infty}(\cup\{x: x \in A_i\})$ and the components of $\cap_{i=1}^{\infty}(\cup\{x: x \in A_i\})$. Andrews and Rubin proved that if for each i, each element of A_i is a solid torus, then $(E^3/G) \times E^1$ is homeomorphic to E^4. It follows from this result that if G is any toroidal point-like decomposition of E^3, then $(E^3/G) \times E^1$ is homeomorphic to E^4.

By using the result [2] of Andrews and Curtis, Meyer proved [60] that if C is any 3-cell in E^3 such that $\text{Bd}\,C$ is locally polyhedral except on an arc α lying on $\text{Bd}\,C$, then $(E^3/G_C) \times E^1$ is homeomorphic to E^4. He also showed that if $\text{Bd}\,C$ is locally polyhedral except on a disc D lying on $\text{Bd}\,C$, then $(E^3/G_C) \times E^1$ and $(E^3/G_D) \times E^1$ are homeomorphic. A result related to these is established in [13].

In [2], Andrews and Curtis stated the following:

Conjecture. If C is a k-cell in E^n, then $(E^n/G_C) \times E^k$ is homeomorphic to E^{n+k}.

It would be of interest to know whether there is a 3-cell C in E^3 such that $(E^3/G_C) \times E^1$ is topologically distinct from E^4. By a theorem of Meyer's mentioned above, if there is such a 3-cell C in E^3, then $\text{Bd}\,C$ fails to be locally tame on some subset of $\text{Bd}\,C$ not contained in any arc on $\text{Bd}\,C$. A candidate for such a 3-cell is one bounded by a 2-sphere constructed by Bing in [24]. Bing constructed a 2-sphere S in E^3, such that S is locally wild at each point of S and S bounds a 3-cell B. It was shown, however, in [11] that $(E^3/G_B) \times E^1$ is homeomorphic to E^4.

Kwun extended the result [2] of Andrews and Curtis by proving [52] that if α is any arc in E^n and β is any arc in E^m, then $(E^n/G_\alpha) \times (E^m/G_\beta)$ is homeomorphic to E^{n+m}.

Question 18. Is it true that if G is any point-like decomposition of E^3, then $(E^3/G) \times E^1$ is homeomorphic to E^4 ?

Question 19. Suppose G is a point-like decomposition of E^3 with only countably many non-degenerate elements. Is $(E^3/G) \times E^1$ homeomorphic to E^4 ?

Question 20. Suppose G is a monotone decomposition of E^3 into one-point sets and a Cantor set of tame arcs. Is $(E^3/G) \times E^1$ homeomorphic to E^4 ? If each arc of G is a straight-line interval? If G is the "unused example" decomposition of section 8 of [27]?

Question 21. Is the Andrews-Curtis conjecture of [2] valid?

The following question can be regarded as a generalization of the conjecture of [2]. In connection with this question, it may be noted that Gillman has shown [40] that if D is any disc in E^3, D is tame in E^4.

Question 22. Suppose C is a k-cell in E^n that is tame in E^{n+j}. Is $(E^n/G_C) \times E^j$ homeomorphic to E^{n+j} ?

§ 13. Decompositions of E^3 into 3-manifolds

Bing showed in [20] that the dogbone decomposition space is not a 3-manifold. In the case of the decomposition G of section 3 of [25], it follows from the argument to show that E^3/G is not homeomorphic to E^3, that E^3/G is not a 3-manifold. In [25], Bing raised the following question: Does a point-like decomposition of E^3 yield E^3 if it yields a 3-manifold?

It was proved in [5] that if G is a point-like decomposition of S^3 such that G has only countably many non-degenerate elements and S^3/G is a 3-manifold, then S^3/G is homeomorphic to S^3. An analogous result holds for E^3.

Connell proved [33] that if G is a monotone decomposition of E^3 into compact absolute retracts such that E^3/G is a contractible 3-manifold with a triangulation T such that any finite subcomplex of T can be piecewise linearly embedded in E^3, then E^3/G is homeomorphic to E^3. It follows from a result of Smale's [74] that "contractible" can be omitted above, because if G is a monotone decomposition of E^3 into compact absolute retracts, E^3/G is contractible.

In this thesis [68], Price proved that if G is a point-like decomposition of E^3 such that E^3/G is a 3-manifold and each compact subset of E^3/G can be embedded in E^3, then E^3/G is homeomorphic to E^3.

Question 23 (Bing). If G is a point-like decomposition of E^3 such that E^3/G is a 3-manifold, is E^3/G homeomorphic to E^3 ?

The following question, more general than that of Bing's above, was raised in [9].

Question 24. If K is a 3-manifold and G is a cellular decomposition of K such that K/G is a 3-manifold, is K/G homeomorphic to K ?

The following partial solution was established in [9]:

THEOREM. Suppose that K is a 3-manifold, G is a cellular decomposition of K, K/G is a 3-manifold M, and M has a triangulation T such that

(1) no vertex of T belongs to $Cl\ P[H_G]$ and
(2) if σ is any 3-simplex of T, $P^{-1}[\sigma]$ lies in an open 3-cell in K.

Then M is homeomorphic to K.

As corollaries of this result, we have in particular the following two:

COROLLARY 1. If G is a point-like decomposition of E^3 such that E^3/G is a 3-manifold and some open set in E^3/G is disjoint from $P[H_G]$, then E^3/G is homeomorphic to E^3.

COROLLARY 2. If G is a point-like decomposition of E^3 such that E^3/G is a 3-manifold and $P[H_G]$ is contained in a compact 0-dimensional set, then E^3/G is homeomorphic to E^3.

The second corollary is also a consequence of results of [8].

Finney proved the following related result [38]: Let M be a triangulated 3-sphere and let T be a triangulated topological space. If there exists a point-like, simplicial mapping from M onto T, then T is homeomorphic to M. By a point-like map is meant a map f such that if y is any point of T, $f^{-1}[y]$ is point-like in M.

Roberts and Steenrod characterized [70] spaces of monotone decompositions of compact 2-manifolds. The decomposition space involved are closely connected with cactoids.

<u>Question</u> 25. Is there an analogous result for point-like decompositions of compact 3-manifolds? Are the decompositions spaces of point-like decompositions of compact 3-manifolds simply related to those of S^3 ?

It was mentioned in section 9 that if G is a monotone decomposition of S^3 such that S^3/G is a 3-manifold, then S^3/G is simply connected. Both Bing and Moise have converses of this result. Moise showed [62] that if M is a compact connected simply connected 3-manifold, there is a monotone decomposition G of S^3 such that S^3/G is homeomorphic to M and each non-degenerate element of G is a linear graph: G can be required to satisfy additional conditions. For a related result due to Bing, see Theorem 4 of [29].

§14. Some related areas

The study of monotone decompositions of E^3, S^3, and 3-manifolds-with-boundary is of considerable importance in a number of problems in topology. In this section we mention three such related areas.

1. Crumpled cubes. See [54] for references to this area.

2. Equivalent decompositions of E^3. The paper [13] introduces the notion of equivalent decompositions of E^3. This idea is a formalization of a natural process in the study of decompositions.

3. Periodic homeomorphisms of E^3 and S^3. Monotone decompositions of E^3 and S^3 may be used to construct unusual periodic homeomorphisms of E^3 and S^3. See [28] for references.

REFERENCES

[1] P. Alexandroff, Über stetige abbildungen kompakter Raüme, Math. Ann. 96 (1927) 555-571.

[2] J. J. Andrews and M. L. Curtis, n-space modulo an arc, Ann. of Math. 75 (1962) 1-7.

[3] J. J. Andrews and L. Rubin, Some spaces whose product with E^1 is E^4, to appear.

[4] L. Antoine, Sur les voisinages de deux figures homeomorphes, Fund. Math. 5 (1924) 265-287.

[5] S. Armentrout, Upper semi-continuous decompositions of E^3 with at most countably many non-degenerate elements, Ann. of Math. 78 (1963) 605-618.

[6] _____, On upper semi-continuous decompositions of E^3 into tame 3-cells and one-point sets, Notices Amer. Math. Soc. 10 (1963) 461.

[7] _____, A property of a decomposition space described by Bing, Notices Amer. Math. Soc. 11(1964) 369-370.

[8] _____, Decompositions of E^3 with a compact 0-dimensional set of non-degenerate elements, to appear.

[9] _____, Concerning cellular decompositions of 3-manifolds that yield 3-manifolds, to appear.

[10] _____, On embedding decomposition spaces of E^n in E^{n+1}, to appear.

[11] _____, Concerning a wild 3-cell described by Bing, to appear.

[12] S. Armentrout and R. H. Bing, A toroidal decomposition of E^3, to appear.

[13] S. Armentrout, L. L. Lininger, and D. V. Meyer, Equivalent decompositions of E^3, these proceedings.

[14] E. Artin and R. H. Fox, Some wild cells and spheres in three-dimensional space, Ann. of Math. 49 (1948) 979-990.

[15] R. J. Bean, Decompositions of E^3 which yield E^3, to appear.

[16] _____, Decompositions of E^3 with a null sequence of starlike equivalent non-degenerate elements are E^3, Notices Amer. Math. Soc. 11 (1964) 558.

[17] R. H. Bing, A homeomorphism between the 3-sphere and the sum of two solid horned spheres, Ann. of Math. 56 (1952) 354-362.

[18] _____, Some monotone decompositions of a cube, Ann. of Math. 61 (1955) 279-288.

[19] _____, Upper semicontinuous decompositions of E^3, Ann. of Math. 65 (1957) 363-374.

[20] _____, A decomposition of E^3 into points and tame arcs such that the decomposition space is topologically different from E^3, Ann. of Math. 65 (1957) 484-500.

[21] _____, Conditions under which monotone decompositions of E^3 are simply connected, Bull. Amer. Math. Soc. 63 (1957) 143.

[22] _____, Necessary and sufficient conditions that a 3-manifold be S^3, Ann. of Math. 68 (1958) 17-38.

[23] _____, The cartesian product of a certain non-manifold and a line is E^4, Ann. of Math. 70 (1959) 399-412.

[24] _____, A wild surface each of whose arcs is tame, Duke Math. J. 28 (1961) 1-15.

[25] _____, Point-like decompositions of E^3, Fund. Math. 50 (1962) 431-453.

[26] _____, Tame Cantor sets in E^3, Pacific. J. Math 11 (1961) 435-446.

[27] _____, Decompositions of E^3, Topology of 3-manifolds and Related Topics, Prentice-Hall (1962) 5-21.

[28] _____, Inequivalent families of periodic homeomorphisms of E^3, Ann. of Math. 80 (1964) 78-93.

[29] _____, Mapping a 3-sphere onto a homotopy 3-sphere, these proceedings.

[30] R. H. Bing and M. L. Curtis, Imbedding decompositions of E^3 in E^4, Proc. Amer. Math. Soc. 11 (1960) 149-155.

[31] L. O. Cannon, Another property that distinguishes Bing's dogbone space from E^3, Notices Amer. Math. Soc. 12 (1965) 363.

[32] B. G. Casler, On the sum of two solid Alexander horned spheres, Notices Amer. Math. Soc. 9 (1962) 108.

[33] E. H. Connell, Images of E_n under acyclic maps, Amer. J. Math. 83 (1961) 787-790.

[34] M. L. Curtis, An imbedding theorem, Duke Math. J. 24 (1957) 349-351.

[35] M. L. Curtis and R. L. Wilder, The existence of certain types of manifolds, Trans. Amer. Math. Soc. 91 (1959) 152-160.

[36] E. Dyer, Certain transformations which lower dimension, Ann. of Math. 63 (1955) 15-19.

[37] E. Dyer and M. E. Hamstrom, Completely regular mappings, Fund. Math. 45 (1958) 103-118.

[38] R. Finney, Point-like, simplicial mappings of a 3-sphere, Canad. J. of Math. 15 (1963) 591-604.

[39] M. K. Fort, Jr., A note concerning a decomposition space defined by Bing, Ann. of Math 65 (1957) 501-504.

[40] D. S. Gillman, Unknotting 2-manifolds in 3-hyperplanes of E^4, to appear.

[41] D. S. Gillman and J. M. Martin, Countable decompositions of E^3 into points and point-like arcs, Notices Amer. Math. Soc. 10 (1963) 74.

[42] R. P. Goblirsch, On decompositions of 3-space by linkages, Proc. Amer. Math. Soc. 10 (1959) 728-730.

[43] O. G. Harold, Jr., A sufficient condition that a monotone image of the three-sphere be a topological three-sphere, Proc. Amer. Math. Soc. 9 (1958) 846-850.

[44] J. G. Hocking and G. S. Young, Topology, Addison-Wesley, 1961.

[45] W. Hurewicz, Über oberhalf-stetige Zerlegungen von Punktmengen in Kontinua, Fund. Math. 15 (1930) 57-60.

[46] L. V. Keldys, Imbedding into E^4 of some monotone images of E^3, Soviet Math. Dokl. 2 (1961) 9-12.

[47] _____, On imbedding certain monotone images of E^n in E^n and E^{n+1} (Russian), Matem. Sb. 57 (1962) 95-104.

[48] _____, Some problems of topology in Euclidean spaces, Russian Mathematical Surveys, 16 (1961) 1-15.

[49] _____, Some theorems on topological imbedding (Russian), General Topology and its Relations to Modern Analysis and Algebra (Proc. Sympos., Prague, 1961), Academic Press (1962) 230-234.

[50] C. Kuratowski, Sur les decompositions semi-continues d'espaces métriques compacts, Fund. Math. 11(1928) 169-185.

[51] K. W. Kwun, Upper semi-continuous decompositions of the n-sphere, Proc. Amer. Math. Soc. 13 (1962) 284-290.

[52] _____, Product of euclidean spaces modulo an arc, Ann. of Math. 79 (1964) 104-107.

[53] K. W. Kwun and F. Raymond, Almost acyclic maps of manifolds, Amer. J. Math. 86 (1964) 638-650.

[54] L. L. Lininger, Crumpled cubes, these proceedings.

[55] J. M. Martin, Sewing of Crumpled cubes which do not yield S^3, these proceedings.

[56] L. F. McAuley, Some upper semi-continuous decompositions of E^3 into E^3, Ann. of Math. 73 (1961) 437-457.

[57] _____, Upper semi-continuous decompositions of E^3 into E^3 and generalizations to metric spaces, Topology of 3-manifolds and Related Topics, Prentice Hall (1962) 21-26.

[58] _____, More point-like decompositions of E^3, Notices Amer. Math. Soc. 11 (1964) 460.

[59] D. R. McMillan, Jr., A criterion for cellularity in a manifold, Ann. of Math. 79 (1964) 327-337.

[60] D. V. Meyer, E^3 modulo a 3-cell, Pacific J. Math. 13(1963) 193-196.

[61] _____, A decomposition of E^3 into points and a null family of tame 3-cells in E^3, Ann. of Math. 78 (1963) 600-604.

[62] E. Moise, Simply connected 3-manifolds, Topology of 3-manifolds and Related Topics, Prentice-Hall (1962) 196-197.

[63] R. L. Moore, Concerning upper semi-continuous collections of continua which do not separate a given set, Proc. Nat. Acad. Sci. 10 (1924) 356-360.

[64] _____, Concerning upper semi-continuous collections of continua, Trans. Amer. Math. Soc. 27 (1925) 416-428.

[65] _____, Concerning upper semi-continuous collections, Monatsh. Math. Phys. 36 (1929) 81-88.

[66] _____, Foundations of Point Set Theory, Amer. Math. Soc. Colloquium Publications 13 (1932).

[67] _____, Foundations of Point Set Theory, (rev. ed.), Amer. Math. Soc. Colloquium Publications 13 (1962).

[68] T. M. Price, Upper semi-continuous decompositions of E^3, Thesis, University of Wisconsin (1964).

[69] J. H. Roberts, Collections filling a plane, Duke Math. J. 2 (1936) 10-19.

[70] J. H. Roberts and N. E. Steenrod, Monotone transformations of two dimensional manifolds, Ann. of Math. 39 (1938) 851-862.

[71] R. H. Rosen, Decomposing 3-space into circles and points, Proc. Amer. Math. Soc. 11 (1960) 918-928.

[72] _____, E^4 is the product of a totally non-euclidean space and E^1, Ann. of Math. 73 (1961) 349-361.

[73] L. Rubin, The product of an unusual decomposition space with a line is E^4, to appear.

[74] S. Smale, A Vietoris mapping theorem for homotopy, Proc. Amer. Math. Soc. 8 (1957) 604-610.

[75] W. R. Smythe, Jr., A theorem on upper semi-continuous decompositions, Duke Math. J. 22 (1955) 485-495.

[76] D. G. Stewart, Cellular subsets of the 3-sphere, Trans. Amer. Math. Soc. 114 (1965) 10-22.

[77] University of Georgia, Summer Institute in Topology, Notes, 1961.

[78] J. J. Wardwell, Continuous transformations preserving all topological properties, Amer. J. Math. 58 (1936) 709-726.

[79] G. T. Whyburn, Non-alternating transformations, Amer. J. Math. 56 (1934) 294-302.

[80] _____, On the structure of continua, Bull. Amer. Math. Soc. 42 (1936) 49-73.

[81] _____, Analytic Topology, Amer. Math. Soc. Colloquium Publications 28 (1942).

[82] _____, Decomposition spaces, Topology of 3-manifolds and Related Topics, Prentice-Hall (1962) 2-4.

[83] R. L. Wilder, Topology of Manifolds, Amer. Math. Soc. Colloquium Publications 32 (1949).

[84] _____, Monotone mappings of manifolds, Pacific J. Math. 7 (1957) 1519-1528.

EQUIVALENT DECOMPOSITIONS OF E^3

by

Steve Armentrout*
Lloyd L. Lininger
Donald V. Meyer**

§1. Introduction

The purpose of this paper is to introduce and study a notion of equivalence of certain types of monotone decompositions of E^3. The notion of equivalence that we study is more restrictive than topological equivalence of the associated decomposition spaces. Detailed proofs will appear elsewhere.

If G is an upper semi-continuous decomposition of E^3, H_G will denote the union of all the non-degenerate elements of G and P_G will denote the projection map from E^3 onto the decomposition space E^3/G associated with G. Suppose that F and G are monotone decompositions of E^3 such that each of $Cl P_F[H_F]$ and $Cl P_G[H_G]$ is compact and 0-dimensional. Then F and G are equivalent decompositions of E^3 if and only if there is a homeomorphism f from E^3/F onto E^3/G such that $f[Cl P_F[H_F]] = Cl P_G[H_G]$.

By introducing the idea of equivalent decompositions of E^3, we are able to analyze, in a precise way, a process that seems quite natural in the study of monotone decompositions of E^3 of the type we are considering. If F is a monotone decomposition of E^3, the stipulation that $Cl P_F[H_F]$ be a compact 0-dimensional set is equivalent to the following condition: There is a sequence $M_1, M_2, M_3, \ldots$ of compact 3-manifolds-with-boundary in E^3 such that for each positive integer j, $M_{j+1} \subset \text{Int } M_j$ and g is a non-degenerate element of F if and only if g is a non-degenerate component of $\cap_{j=1}^{\infty} M_j$.

A process one finds useful in certain situations is one that involves a sequence $f_1, f_2, f_3, \ldots$ of homeomorphisms from E^3 onto E^3 such that

(1) f_1 shrinks or stretches M_1,
(2) f_2 agrees with f_1 on $E^3 - M_1$ and shrinks or stretches M_2,

* Supported by NSF Grant No. GP-4508
** Supported by an NSF Academic Year Extension.

(3) f_3 agrees with f_2 on $E^3 - M_2$ and shrinks or stretches M_3, and so on. The "new" decomposition has as its non-degenerate elements the non-degenerate components of $f_1[M_1] \cap f_2[M_2] \cap f_3[M_3] \cap \dots$. We are able to show that under fairly mild restrictions, there exists such a sequence of homeomorphisms if and only if the original decomposition and the "new" one are equivalent in the sense of this paper.

We indicate some examples that illustrate these concepts. The first two examples give instances of previous applications of the ideas of this paper. The remaining ones are described in detail in the present paper.

Example 1. Meyer proved [7] that if C is a 3-cell in E^3 such that Bd C is locally polyhedral except at points of an arc α on Bd C, then E^3/C is homeomorphic to E^3/α. His proof shows that the decomposition of E^3 whose only non-degenerate element is C and that whose only non-degenerate element is α are equivalent.

Example 2. Bing described [4] a 2-sphere S in E^3 such that S is locally wild at each point of S and S bounds a 3-cell B in E^3. Armentrout proved [1] that there is a 3-cell B' in E^3 such that Bd B' is locally polyhedral except on a Cantor set on Bd B' and E^3/B is homeomorphic to E^3/B'. The argument of [1] establishes that the decomposition of E^3 whose only non-degenerate element is B and that whose only non-degenerate element is B' are equivalent.

Example 3. Suppose G is a monotone decomposition of E^3 such that there is a sequence $M_1, M_2, M_3, \dots$ of compact 3-manifolds-with-boundary as described above. Suppose further that each component of each M_i is a 3-cell-with-handles. Then G is equivalent to a decomposition into 1-dimensional continua and one-point sets; see section 5.

Example 4. Bing's dogbone decomposition [3] is equivalent to a decomposition into one-point sets and, at most, countably many non-degenerate continua; see section 3.

Example 5. In section 3 of [5], Bing described a point-like decomposition G of E^3 with only countably many non-degenerate elements such that E^3/G is not homeomorphic to E^3. There exists a decomposition F of E^3 such that F is equivalent to G and F has uncountably many non-degenerate elements; see section 3.

The notation and terminology of [2] will be followed.

§ 2. The existence of sequences of homeomorphisms

A compact continuum M in E^3 is semi-cellular if and only if for each open set U in E^3 containing M, there is an open set V lying in U and containing M and such that each simple closed curve in V is null-homotopic in U. Each point-like compact continuum in E^3 is semi-cellular, and so is each compact absolute retract.

THEOREM 1. Suppose that F and G are monotone decompositions of E^3 such that $C\ell P_F[H_F]$ and $C\ell P_G[H_G]$ are compact 0-dimensional sets. Suppose that F is definable by 3-cells-with-handles. Suppose each element of G is semi-cellular. Then if F and G are equivalent decompositions there exists a sequence $f_1, f_2, f_3, \ldots$ of homeomorphisms from E^3 onto E^3 such that

(1) for each i, $f_{i+1}|(E^3 - \text{Int } M_i) = f_i|(E^3 - \text{Int } M_i)$, and

(2) $f_1[M_1]$, $f_2[M_2]$, $f_3[M_3], \ldots$ is a defining sequence for G.

COROLLARY 1. If F and G satisfy the hypothesis of Theorem 1, then G is definable by 3-cells-with-handles. If F is toroidal, so is G.

THEOREM 2. Suppose that F and G are monotone decompositions of E^3 such that $C\ell P_F[H_F]$ and $C\ell P_G[H_G]$ are compact 0-dimensional sets. Suppose that F has a defining sequence $M_1, M_2, M_3, \ldots$ and there exists a sequence $f_1, f_2, f_3, \ldots$ of homeomorphisms from E^3 onto E^3 such that

(1) for each i, $f_{i+1}|(E^3 - \text{Int } M_i) = f_i|(E^3 - \text{Int } M_i)$, and

(2) $f_1[M_1]$, $f_2[M_2]$, $f_3[M_3], \ldots$ is a defining sequence for G.

Then F and G are equivalent.

THEOREM 3. Suppose F and G are monotone decompositions of E^3 such that $C\ell P_F[H_F]$ and $C\ell P_G[H_G]$ are compact 0-dimensional sets. Suppose F is definable by 3-cells-with-handles and each element of G is semi-cellular. Then F and G are equivalent if and only if there exists a defining sequence $M_1, M_2, M_3, \ldots$ for F such that for each positive integer i, each component of M_i is a 3-cell-with-handles and a sequence $f_1, f_2, f_3, \ldots$ of homeomorphisms from E^3 onto E^3 such that

(1) for each i, $f_{i+1}|(E^3 - \text{Int } M_i) = f_i|(E^3 - \text{Int } M_i)$, and

(2) $f_1[M_1]$, $f_2[M_2]$, $f_3[M_3], \ldots$ is a defining sequence for G.

See [2] for some conditions on a monotone decomposition F under which F is definable by 3-cells-with-handles.

§3. Results on the dogbone and point-like decompositions of Bing

THEOREM 4. There exists a point-like decomposition F equivalent to the dogbone which has only countably many non-degenerate elements.

THEOREM 5. If F is a point-like decomposition equivalent to the point-like decomposition G of section 3, [5], then some non-degenerate element of F is not locally connected.

COROLLARY 2. There does not exist a point-like decomposition F equivalent to G such that each non-degenerate element is an arc.

COROLLARY 3. The decomposition G is not equivalent to the dogbone decomposition.

Example 6. There exists a point-like decomposition F such that if K is any point-like decomposition equivalent to F then K has uncountably many non-degenerate elements.

§4. Tamely finnable 3-cells

A 3-cell C in E^3 is tamely finnable if and only if there exists a tame disc D in E^3 such that $D \cap C$ is an arc α and $\alpha \subset (\text{Bd } D) \cap (\text{Bd } C)$.

THEOREM 6. Let C be a 3-cell in E^3 such that C is tamely finnable. Then there exists a 3-cell C' in E^3 such that C' has a flat spot and the decomposition of E^3 whose only non-degenerate element is C is equivalent to the decomposition of E^3 whose only non-degenerate element is C'.

The proof consists of pushing apart the boundary of the 3-cell at the intersection of the fin and the boundary of the 3-cell, and then inserting a flat disc.

COROLLARY 4. If C is a 3-cell in E^3 and C is tamely finnable, then there exists a disc D in E^3 such that the decomposition of E^3 whose only non-degenerate element is C is equivalent to the decomposition of E^3 whose only non-degenerate element is D.

This follows from Theorem 3 of [7].

THEOREM 7. If K is a crumpled cube in E^3, there is a 3-cell C in E^3 such that the decomposition of E^3 whose only non-degenerate element is K is equivalent to the decomposition whose only non-degenerate element is C.

Proof. Apply Theorem 2 of [6].

Question 1. Does there exist a 3-cell C in E^3 such that for each disc D in E^3, E^3/C and E^3/D are not equivalent?

§5. Improving elements of decompositions

THEOREM 8. Suppose F is an upper semi-continuous decomposition of E^3 and F is definable by 3-cells-with-handles. Then there exists an upper semi-continuous decomposition G of E^3 such that F is equivalent to G and each non-degenerate element of G is 1-dimensional.

The proof consists of repeatedly pulling the 3-cells-with-handles toward certain 1-dimensional spines.

THEOREM 9. Let G be a monotone upper semi-continuous decomposition of E^3 such that G has only countably many non-degenerate elements; each non-degenerate element is tame (relative to the usual triangulation of E^3), and H_G is bounded. Then there exists a homeomorphism h of E^3 onto itself such that if $g \in G$, $h[g]$ is polyhedral.

COROLLARY 5. If F is an upper semi-continuous decomposition of E^3 into tame 3-cells and one-point sets such that H_F is bounded, then there exists an upper semi-continuous decomposition G of E^3 into polyhedral 3-cells and one-point sets such that F is equivalent to G.

Question 2. Suppose F is an upper semi-continuous decomposition of E^3 into one-point sets and 3-cells (perhaps tame). Does there exist an upper semi-continuous decomposition G of E^3 into one-point sets and arcs (or trees) such that F and G are equivalent?

REFERENCES

[1] S. Armentrout, Concerning a wild 3-cell described by Bing, to appear.

[2] _____, Monotone decompositions of E^3, these proceedings.

[3] R. H. Bing, A decomposition of E^3 into points and tame arcs such that the decomposition space is topologically different from E^3, Ann. of Math. 65 (1957) 484-500.

[4] _____, A wild surface each of whose arcs is tame, Duke Math. J. 28 (1961) 1-15.

[5] _____, Point-like decompositions of E^3, Fund. Math. 50 (1962) 431-453.

[6] L. L. Lininger, Some results on crumpled cubes, Trans Amer. Math. Soc. 118 (1965) 534-549.

[7] D. V. Meyer, E^3 modulo a 3-cell, Pacific J. Math. 13 (1963) 193-196.

ANOTHER DECOMPOSITION OF E^3 INTO POINTS AND INTERVALS

by

Louis F. McAuley

An intensive effort has been made in recent years to find some satisfactory analogue for E^3 to Moore's theorem concerning point-like decompositions of a plane E^2. And, there seems to be no such comparable result. The topology of E^3 is incredibly complicated relative to the topology of E^2.

Various researchers have provided useful information about the topology of E^3. We do not attempt to list all such names. Yet, we must mention the names of Bing and Moise who have given us many of our tools.

Let us restrict ourselves to the following problem which has been the subject of numerous papers in the past decade.

Problem. Suppose that G is an upper semi-continuous decomposition of E^3. Under what conditions is the decomposition space homeomorphic to E^3 ?

There is no satisfactory answer to this question although conditions for an affirmative answer are given in [5] and even more general conditions by McAuley in [13, 14], but Bing's work has set the stage for various research on this problem.

It may be worthwhile to enumerate both some affirmative and some negative results. First, let us consider some basic definitions and terminology.

Suppose that G is a collection of pairwise disjoint compact continua filling up a metric space M (that is, a decomposition of M). Furthermore, each element g of G is point-like, that is, $M - g$ is homeomorphic to the complement of a point in M. The letter G shall also denote the decomposition space associated with G where topology is defined by the various sub-collections R of G which are designated as open sets in G. We say that $R \subset G$ is open in G if and only if R^* (the union of the elements of R) is open in M. We further require G to be upper semi-continuous. Thus, if $A \subset G$, g and h are two limit elements of A, then there exists $B \subset A$ such that g is a limit element of B but h is not a limit element of B (Hausdorff Property).

If G is an upper semi-continuous decomposition of E^3 (called simply a decomposition of E^3 for the purposes of this paper), then the decomposition space may be topologically E^3 even though the elements of G are not point-like. See the remarkable example in [8] due to Bing.

Let us suppose that G is always an u.s.c. decomposition of E^3 and that H is the collection of non-degenerate elements of G.

Affirmative Answers.

(1) H is countable and H^* is a G_δ set. Bing, 1957, [5].
(2) Elements of H are starlike. Bing, 1957, [5].
(3) Each element of H lies in a horizontal plane. Dyer and Hamstrom, 1958, [9].
(4) There is a countable number of planes such that each element of H is perpendicular to one of these planes. McAuley, 1961 [13].
(5) H is a countable collection of almost tame arcs. Gillman and Martin, 1961, [11].

Negative Answers.

(1) H consists of tame arcs. Bing, 1957, [4].
(2) H is a countable collection of point-like continua. Bing, 1962, [7].

In 1961, Bing published an example in [8] and also [7] in 1962, of a decomposition G of E^3 such that each element of H is a straight line interval. The decomposition space appears to be different from E^3 and Bing points out some properties of his example to indicate that his conjecture is correct. We give in this paper another example of a decomposition G of E^3 into straight line intervals and points. Also, we conjecture that the decomposition space is not homeomorphic to E^3. Our construction is somewhat less complicated than Bing's in [7] but the resulting decomposition does not seem entirely unrelated to his.

It should be apparent that there is a great need for new tools and techniques in studying decompositions of E^3.

We deeply appreciate the stimulating conversations held with R. H. Bing concerning these problems. We are also indebted to Lawrence Cannon and C. E. Burgess for helpful criticisms.

Description of the example. In an effort to simplify a rather complicated construction, we first consider a method for building a certain collection of straight line intervals in a rectangle. These will be used as a kind of one-dimensional approximation to some "straight" slender 3-cells with very small handles. A careful study of the various figures should assist the reader in understanding the construction.

(1) Selecting intervals inside rectangles. Suppose that ABCD is a rectangle with two parallel sides denoted by AB and DC. Let A_1B_1 and D_1C_1 denote intervals parallel to AB and DC, respectively, spanning ABCD. In our construction, these will be close to AB and DC. Also, let EF denote a perpendicular bisector of AB at E and DC at F. Now, $EF \cap D_1C_1 = G$. Next, let D_2C_2 be an interval parallel to DC which separates D_1C_1 from DC in ABCD. Suppose that P and Q are points on A_1B_1 such that EF separates P from Q in ABCD and the intervals A_1P and QB_1 have the same length. Now, let QU and PW denote intervals which intersect at G such that U and W lie on D_2C_2. Construct intervals PU and QW. Denote $EF \cap D_2C_2$ by V and consider intervals PV and QV. Consult Figure 1.

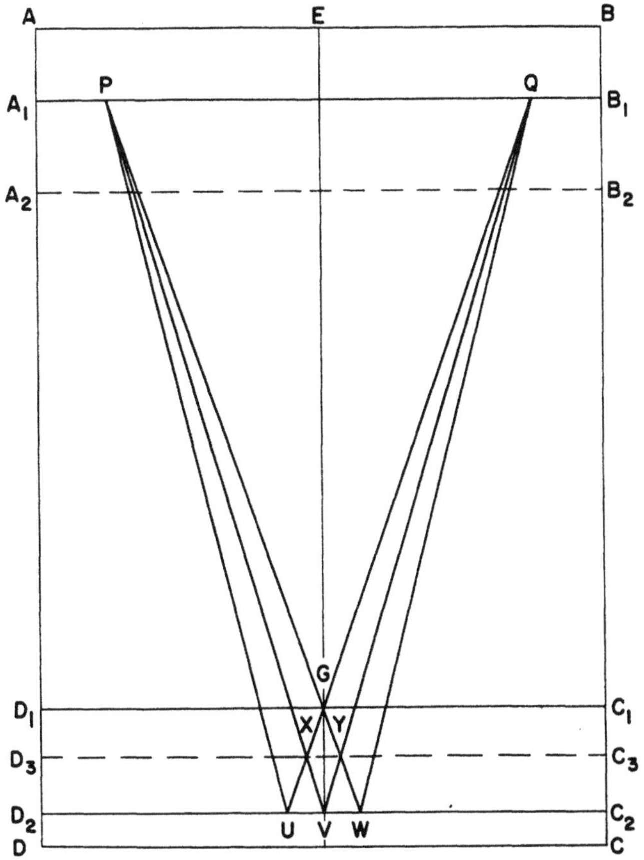

Fig. 1

Note that the points $QU \cap PV = X$ and $PW \cap QV = Y$ lie on an interval D_3C_3 parallel to DC and, of course, separating D_1C_1 from D_2C_2 in ABCD. We shall indicate how to cut out holes along the interval D_3C_3 so that the holes are larger than the gaps between them. In the same manner, we cut holes along an interval A_2B_2 parallel to AB but close to A_1B_1 such that A_1B_1 separates AB from A_2B_2 in ABCD. These facts are easily observed from Figure 2.

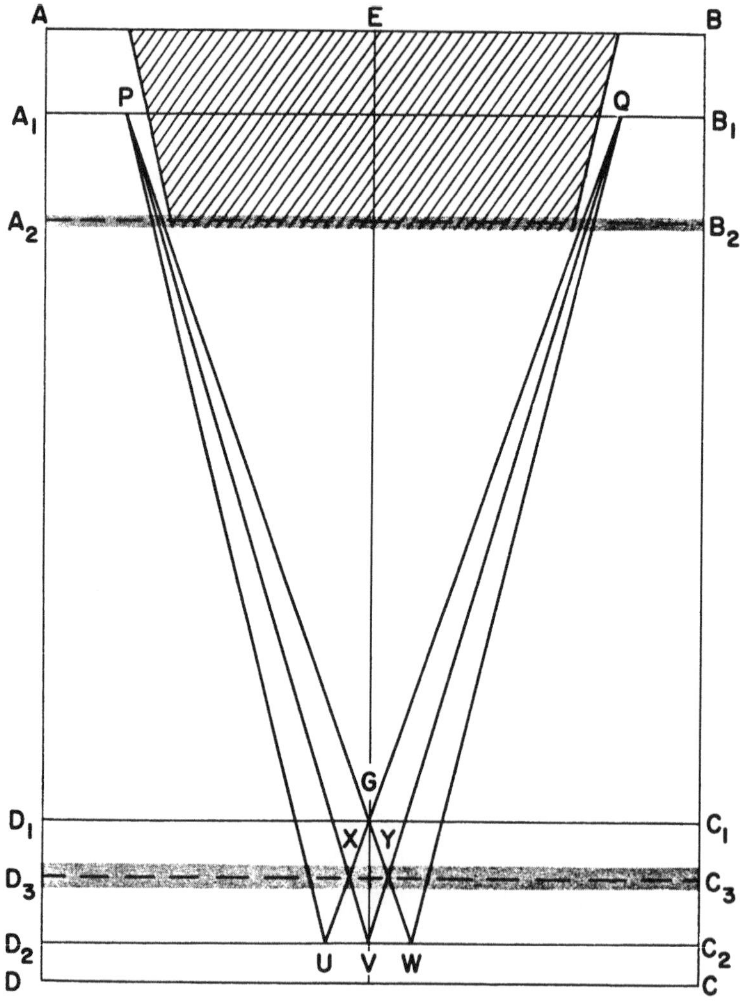

Fig. 2.

We can see that we have two types of cuts indicated by the shaded areas in Figure 2. One type is a slot which removed part of the original boundary of ABCD while the other type is a hole in the rectangle ABCD. And, it should be clear that the holes can be made as narrow as we please so that the gaps in between the holes can be made as small as we please. Denote this special punctured rectange by ABCD.

(2) Adjusting the six intervals—a pair of triods—inside a cube with handles. In (1), we constructed a rectangle ABCD with slots and holes. Let this object be called R. Now, consider $R\times[0, t]$ for $t > 0$. This is a cube with handles. The section $R\times(t/2)$ looks like Figure 2. Let us rotate the left triod LT consisting of the intervals PU, PV, and PW slightly about the interval D_2C_2 in one direction and rotate the right triod RT consisting of QU, QV, and QW slightly in the other direction about D_2C_2.

Now, we have a cube with four holes at one end and three at the other. Inside this object, we have two triods LT and RT of straight line intervals which intersect it only at their endpoints U, V, and W. Furthermore, these triods form a theta curve which links around the seven holes in this cube. Let us call this special cube an R_7-cube. Next, we wish to replace each interval of each of the two triods in this R_7-cube with thin R_7-cubes which link through these seven holes. We shall do this in stages. First, we expand the intervals to rectangles.

(3) Expanding intervals in the triods LT and RT into rectangles. Let us suppose that we have an R_7-cube with triods LT and RT as constructed above. Now, inside A, we wish to expand these triods to triods of rectangles intersecting in a special way. In fact, in planes perpendicular to the "top" face of A, we take small rectangular nbhds of each interval in LT and RT which preserves their intersection properties. We indicate this in Figure 3.

Our triods LT and RT consist now of rectangles instead of intervals. These intersect along intervals P, Q, U, V and W instead of at points P, Q, U, V, and W.

(4) Cutting slots and holes along intervals P, Q, U, V, and W in the rectangles of LT and RT so that these become copies of the punctured rectangles ABCD. In (1), we constructed a special punctured rectangle ABCD with a pair of triods. Now, we wish to replace each rectangle in each of the triods of rectangles LT and RT by copies of the punctured rectangle ABCD of (1) which link each other through the holes along the intervals P, Q, U, V, and W. It is easy to see how this may be accomplished by studying Figures 3, 4, and 5. In Figure 4, we show only two of the punctured rectangles linking but the third links each of these two in exactly the same way.

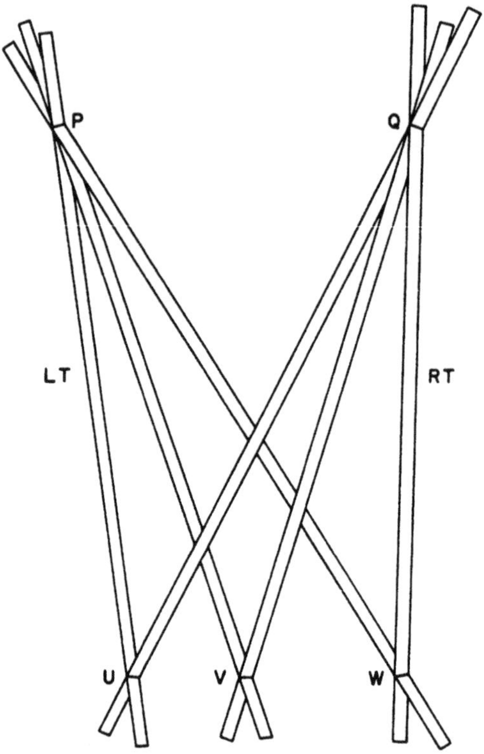

Fig. 3

It is essential that the holes and slots be cut in such a way that we have punctured rectangles ABCD (Figure 2) which contain a pair of triods of intervals as constructed in (1). We can use the pairs of intervals (P, U), (P, V), (P, W), (Q, U), (Q, V), and (Q, W) as the intervals A_2B_2 and D_3C_3, respectively, in Figure 1 as guide lines for cutting the various holes and slots.

(5) <u>Thickening the triods of disjoint linking punctured rectangles into corresponding triods of R_7-cubes</u>. We now have six punctured rectangles R_i linking in the way we described above inside an R_7-cube A. Now, we select t, $0 < t < \frac{1}{2}$, such that the collection of 2t-nbhds $N_{2t}(R_i)$ of the various R_i are pairwise disjoint. Inside, each nbhd, we take $R_i \times [0,t]$ which is an R_7-cube. Thus, we now have in the interior of A, denoted by Int A, six R_7-cubes $A_1, A_2, \ldots, A_6$ which are pairwise disjoint but link in the special way described above for the sections $R_i \times 0$. For each section $R_i \times t/2$, we have a copy of one punctured rectangle R

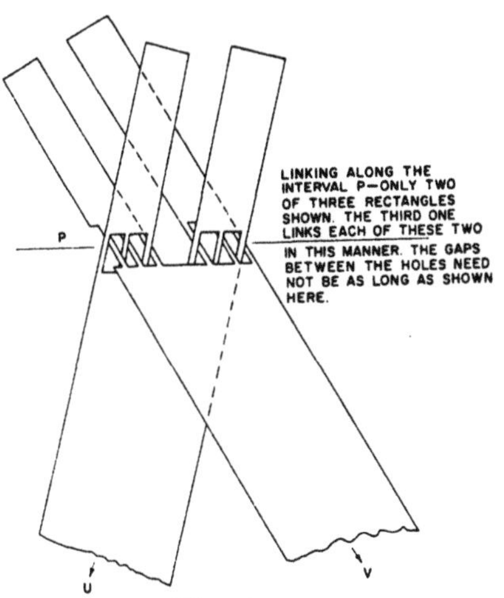

Fig. 4

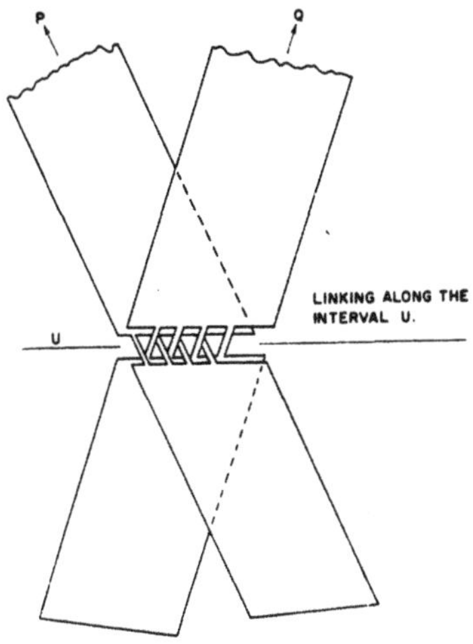

Fig. 5

in which we have a pair of triods LT and RT which link around the holes as prescribed in (1).

(6) We repeat the construction described in steps (2)-(5) using instead of A each of the R_7-cubes A_{1i}, $i = 1,2,\dots,6$. In step (5), we choose t, $0 < t < \frac{1}{4}$. Thus, for each i, we obtain six R_7-cubes A_{i1}, $A_{i2},\dots, A_{i6}$, in Int A_i which have the same linking properties with respect to A_i as the six R_7-cubes $A_1, A_2,\dots, A_6$ have with respect to A. In general, for each $j > 1$, we construct in Int $A_{i\dots j}$ six R_7-cubes $A_{i\dots jk}$, $k = 1,2,\dots,6$, as outlined in steps (2)-(5) for A but choose t, $0 < t < \frac{1}{2}^j$, in step (5). These R_7-cubes have the same linking properties with respect to $A_{i\dots j}$ as the six R_7-cubes $A_{i\dots j-1,n}$, $n=1,2,\dots,6$, have the respect to $A_{i\dots j-1}$.

The Example: In the description above, we have indicated a construction for what we call an R_7-cube A. In its interior, we have described other R_7-cubes, and so forth. For each i, let A_i denote the collection of all R_7-cubes $A_{n_1 n_2 \cdots n_i}$ where $n_j = 1,2,3,\dots,6$. For $i = 1$, we have a collection A_1 which contains an element A_1 but no confusion should result. For $i > 1$, the multiple subscripts should prevent any confusion in this regard.

Consider the point set

$$M = \bigcap_{i=1}^{\infty} A_i^* \quad .$$

where A_i^* denotes the union of the elements of A_i. Each component of M is a straight line interval. Now, let H be the collection of these components. And, let $G = H$ plus those points of $E^3 - H^*$. Thus, G is an upper semi-continuous (u.s.c.) decomposition of E^3. In the decomposition space G, H is homeomorphic to a Cantor set.

<u>Properties of</u> A <u>and the collections</u> A_1. We shall use numerous fundamental theorems about the topology of E^3 from papers of R. H. Bing and E. E. Moise. Results from the paper [4] of Bing will be used extensively, and therefore, an attempt will be made to make definitions and use terminology consistent with his paper.

Topologically, we may consider A as some ε-nbhd of a linear graph F which lies in a plane and consists of seven simple closed curves U_i and L_K, $i=1,2,3,4$ and $k=1,2,3$ such that each pair of different simple closed curves intersects in a point c. See Figure 6. We shall call F the <u>center of</u> A, and the simple closed curves in F are the <u>loops of</u> F.

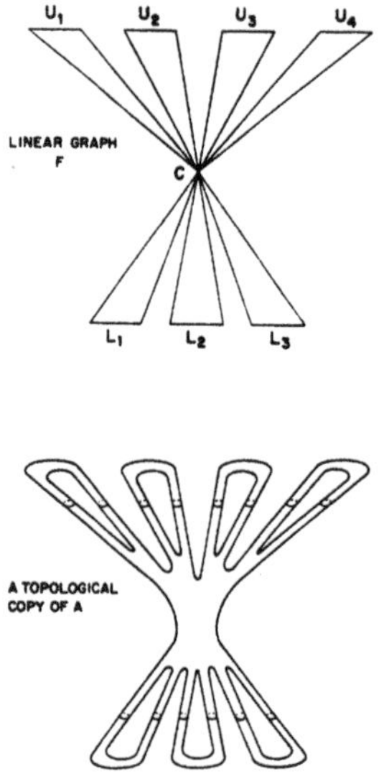

Fig. 6

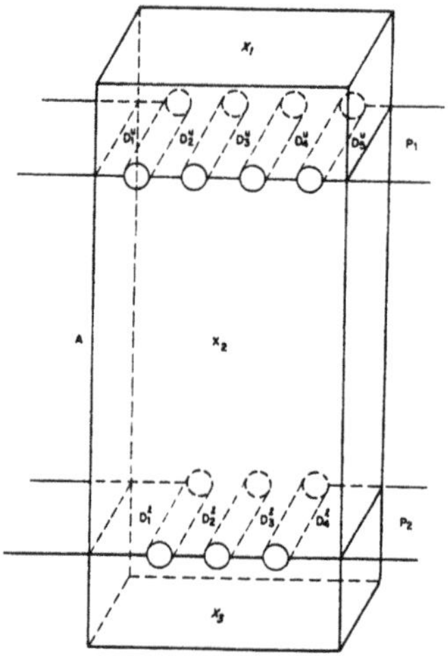

Fig. 7

Now let P_1 and P_2 denote two disjoint parallel planes such that $P_1 \cap A$ = five planar discs which are pairwise disjoint and $P_2 \cap A$ = four planar discs which are pairwise disjoint (i.e., P_1 cuts through the "top" holes of A and P_2 cuts through the "bottom" holes). See Figure 7.

We indicate in Figure 8 a topological picture of the six R_7-cubes A_1, how they link around each other, and how they link around the holes in A.

If B is the image of A under a homeomorphism h, we call h(F) the center of B (the center depends on the homeomorphism h). Suppose that F_1 is a topological graph homeomorphic to F, the center of A. We say that F_1 is <u>homotopic to the center of</u> A if there are continuous maps $f(x, t)$, $x \in F$ and $0 \leq t \leq 1$, of $F\times[0, 1]$ into A such that $f(x, 0) = x$ and $f(x, 1)$ takes F homeomorphically onto F_1. Consider now, two properties of certain topological graphs in A.

<u>Property</u> $P(k, n)$. A topological graph F_1 homeomorphic to the center F of A has Property $P(k, n)$ if and only if F_1 contains $k + n$ points p_i, q_j, $i = 1, \ldots, k$ and $j = 1, 2, \ldots, n$ where no two of these points lie in the same loop (simple closed curve) such that each arc from p_i to q_i in F_1 intersects both P_1 and P_2.

<u>Property</u> $Q(k, n)$. An image of A under a homeomorphism h has Property $Q(k, n)$ if and only if each topological graph F_1 in $h(A)$ which is homeomorphic to the center F of A and which is homotopic to the center $h(F)$ of $h(A)$ also has Property $P(k, n)$.

The next objective is to show that if h is a homeomorphism of E^3 onto itself which is fixed outside A, then for each n, there is some element of A_n, say $A_{ij\cdots k}$, such that $h(A_{ij\cdots k})$ intersects both P_1 and P_2. Our plan is to use arguments like those due to Bing in [4]. However, the situation here is much more complicated than in [4] although enough like [4] to use many of the basic tools developed there. Bing defined and used only Properties $P(1, 1)$ and $Q(1, 1)$.

<u>Reasons for seven holes in the elements of</u> A_1. Consider a homeomorphism h of A onto itself fixed on Bd A. Then it is not difficult (although it will be proved) to see that for some i, $i = 1, \ldots, 6$, $h(A_i)$ intersects both P_1 and P_2 where A_i is an element of A_1. However, it <u>may be</u> that $h(A_i)$ has Property $Q(k, n)$ for $k + n < 7$. If no $h(A_i)$ has Property $Q(k, n)$ for $k + n = 7$, then at the next stage, we can find a homeomorphism g of $h(A_i)$ for each i onto itself fixed on Bd $h(A_i)$ such that $gh(A_{ij})$ fails to have even Property $Q(1, 1)$ for any j.

The homeomorphism h may shrink A_i for some i so that $h(A_i)$ misses $P_1 \cup P_2$ altogether, while for some j, $h(A_j)$ has perhaps only one handle which intersects P_1 and two handles which intersect P_2. It can be proved, however, that because of the seven holes and the special linking of the A_i, there will exist m such that either three handles of $h(A_m)$ on one end intersect P_1 while two handles on the other end of $h(A_m)$ intersect P_2 or the other way around. That is, we can establish either $P(3, 2)$ or $P(2, 3)$. If we could prove $P(4, 3)$, then we could show that the decomposit i in space is not homeomorphic to E^3. But, is either $P(3, 2)$ or $p(2, 3)$ sufficient to yield such a result? We prove the following:

THEOREM 1. Suppose that h is a homeomorphism of A onto itself which is fixed on the boundary of A. Then $h(A)$ has Property $Q(4, 3)$.

Proof: If F is the center of A, then $h(F)$ is homotopic in A to F since there is a graph F_1 on $\mathrm{Bd}\, A$ which is homotopic to F in A. By a result of Hamstrom [12], there is an isotopy of E^3 onto itself fixed outside A and pulls h onto the identity.

Let U_i be one of the four "upper" loops of F, the center of A, and tU_i be its image under a homotopy in A. It will follow that tU_i intersects P_1. Now, consider $P_1 \cap A = \cup_{i=1}^{5} D_{1i}$, the union of five disjoint discs. Suppose that U_i links $\mathrm{Bd}\, D_i$ where we use the same concept of linking (mod 2) as given in [4]. It will follow by an argument like the one for Theorem 11 of [4] that tU_i intersects D_i and therefore, P_1. Furthermore, if tL_j is the homotopic image of a "lower" loop of F, then tL_j intersects P_2. Note also that $P_2 \cap A = \cup_{i=1}^{4} D_{2i}$, four disjoint discs. It should now be clear that $h(A)$ has Property $Q(4, 3)$.

In the proof of the following theorem, we use Theorems 1-7 of [4] which are basic facts about E^3. We do not mention these theorems specifically in the argument which follows because our proof parallels that given by Bing in [4] for Theorem 10. Nevertheless, they are essential.

Admissible Sets of Loops. Since B has Property $Q(4, 3)$, each topological graph F_1 in it, homotopic to its center $h(F)$, has Property $P(4, 3)$. Thus, $h(F)$ has Property $P(4, 3)$. Suppose that U_1, U_2, U_3, and U_4 are the four upper loops of $h(F)$ while L_1, L_2, and L_3 are the three lower loops of $h(F)$. Furthermore, there are four points $p_i \in U_i$ and three points $q_j \in L_j$ where

(1) $p_i, q_j \notin c = U_i \cap L_j$ and

(2) each arc $p_i q_j$ in $h(F)$ intersects both P_1 and P_2.

Clearly, $h^{-1}(U_i)$ is not homotopic to $h^{-1}(L_j)$ in A for any i, j. Consequently, U_i is not homotopic to L_j.

For convenience, assume that $h^{-1}(U_i)$ and $h^{-1}(L_j)$ for $i = 1,2,3,4$ and $j = 1,2,3$, are linked in the order around the four top holes a, b, c, d, of A and around the three bottom holes e, f, g of A, respectively, as shown in Figure 9. Note that $h^{-1}(U_i)$ intersects $h^{-1}(P_1)$ while $h^{-1}(L_j)$ intersects $h^{-1}(P_2)$.

Consider the six elements A_i of the collection A_1 which lie in Int A and which are linked as indicated in Figure 8. Let F_i be a topological graph in $h(A_i)$ which is both homotopic in $h(A_i)$ to its center $h(C_i)$ and homeomorphic to the center C_i of A_i. There is no loss of generality in supposing that no one of the the seven loops (simple closed curves) of F_i lies entirely in $(P_1 \cup P_2) \cap A$.

For each i, let U_1^i, U_2^i, U_3^i, U_4^i and L_1^i, L_2^i, L_3^i, denote the four upper and the three lower loops, respectively, of F_i. Since F_i is homotopic in $h(A_i)$ to its center $h(C_i)$, clearly, $h^{-1}(F_i)$ is homotopic in A_i to C_i. As a matter of convenience and without loss of generality, we shall

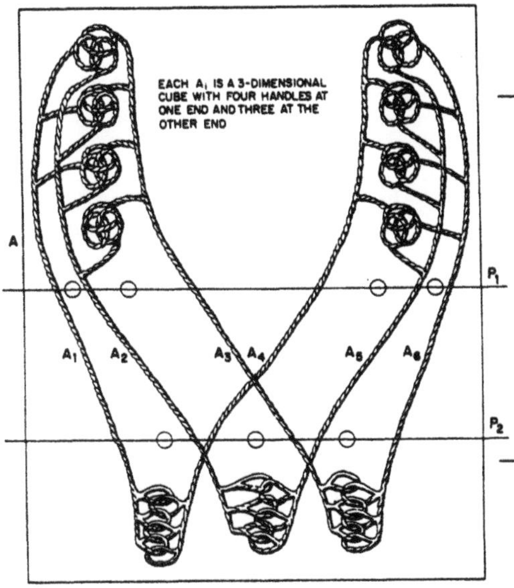

Fig. 8

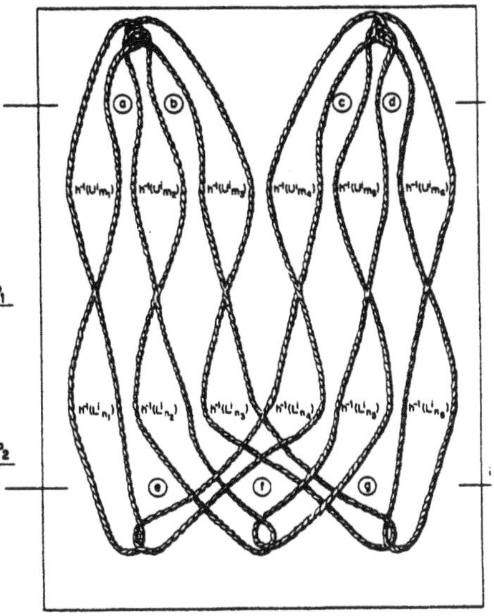

Fig. 9

assume that F_i is $h(C_i)$ and that $h^{-1}(F_i) = C_i$.

From each F_i, we can pick a pair of loops $(U^i_{m_i}, L^i_{n_i})$ such that the six pairs $[h^{-1}(U^i_{m_i}), h^{-1}(L^i_{n_i})]$ link as shown in Figure 10. We shall

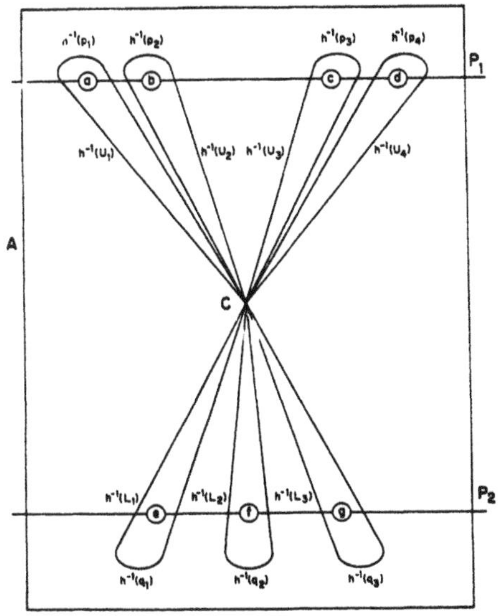

Fig. 10

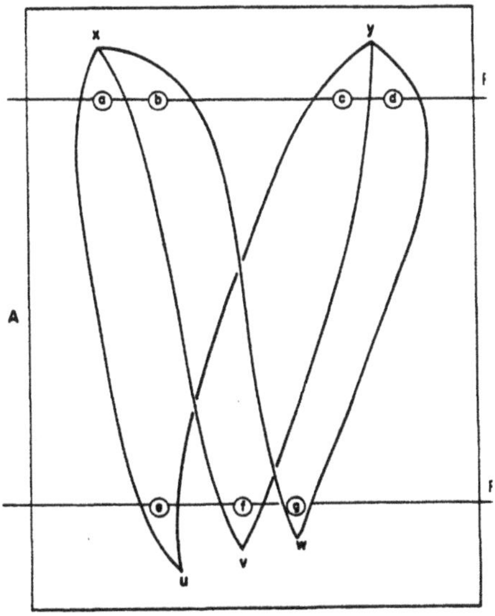

Fig. 11

say that a set of six pairs which link in this manner is <u>an admissible set of</u> loops. The number of such admissible sets of loops is $(4^2) \cdot (3^3) = 432$.

THEOREM 2. If S is an admissible set of loops, then at least one pair $[h^{-1}(U^i_{m_i}), h^{-1}(L^i_{n_i})]$ has the property that $U^i_{m_i} \cup L^i_{n_i}$ intersects both P_1 and P_2.

<u>Proof</u>: Suppose that the theorem is false. There is an $\varepsilon > 0$ so that an ε-nbhd of each pair $[h^{-1}(U^i_{m_i}), h^{-1}(L^i_{n_i})]$ lies in Int A_i. Denote this nbhd by D_i. Thus D_i is a "dogbone" like those used in the paper [4] by Bing. Furthermore, D_i can be selected so that $h(D_i)$ fails to intersect both P_1 and P_2.

Now, we claim that there exist six arcs xu, xv, xw, yu, yv, and yw such that (1) two arcs intersect if they have exactly one point in common which is an endpoint of each, (2) their union is a theta curve, (3) the image under h of each arc fails to intersect both P_1 and P_2, (4) except for "small" nbhds of their endpoints, these arcs lie, respectively, in $A_1, \ldots, A_6$, and (5) these arcs are homotopic to those shown in Figure 11.

The construction of these arcs may be carried out in almost exactly the same way as Bing constructs the four arcs pq_ir_is on pages 493-496 of [4]. Although our example is more complicated, the techniques for producing these arcs are essentially the same. We feel that it is unnecessary to reproduce his long but ingenious argument.

The next objective is to show that the existence of such arcs implies that B does not have Property Q(4, 3).

We have assumed that no one of these arcs intersects both $h^{-1}(P_1)$ and $h^{-1}(P_2)$. Let T denote the theta curve $xu \cup xv \cup xw \cup yu \cup yv \cup yw$. There are exactly three simple closed curves in T. Each of these consists of four arcs—two with x an an endpoint and two with y as an endpoint. See Figure 11. By a theorem of Moise [16] (also, Bing), there is no loss of generality in assuming that h is piecewise linear and that each of the simple closed curves J_{uv}, J_{uw}, and J_{vw} in T is polygonal. The subscripts indicate the points other than x and y which lie on the simple closed curves. Furthermore, we may assume that (1) $h^{-1}(P_1)$ and $h^{-1}(P_2)$ are polygonal planes (which, of course, do not intersect) and (2) each of the three simple closed curves in T is unknotted and bounds a disc. We shall denote these discs by D_{uv}, D_{uw}, and D_{vw}, respectively.

Suppose that $J_{uv} \cap h^{-1}(P_1) = \varphi$. Then we wish to show that D_{uv} can be taken so that $D_{uv} \cap h^{-1}(P_1) = \varphi$. This may done by considering $h(J_{uv})$, a polygonal simple closed curve, which does not intersect P_1. Since J_{uv} is unknotted and h is a homeomorphism of E^3 onto itself, $h(J_{uv})$ is unknotted. Also, we have assumed that h is piecewise linear. Hence, $h(J_{uv})$ bounds a polygonal disc D, i.e., D has a (finite) triangulation T.

Now, D can be adjusted so that no vertex or edge of a triangle lies in P_1. Then each component $D \cap P_1$ is a simple closed curve. Consider the component C of $D - P_1$ which contains J_{uv}. Thus, C is a disc with a finite number of holes bounded by simple closed curves lying in P_1. We now have a polygonal disc D bounded by J_{uv} whose intersection with P_1 is a finite number of pairwise disjoint discs. Next, we adjust D slightly to miss P_1 altogether. Consequently, we may assume that $h(D_{uv})$ and D_{uv} are polygonal discs which miss P_1 and $h^{-1}(P_1)$, respectively.

It is not difficult to see that there is a four leaf clover E_1 in D_{uv} which is homotopic in A to $E_2 = h^{-1}(U_1^i) \cup h^{-1}(U_3^i) \cup h^{-1}(L_1^i) \cup h^{-1}(L_2^i)$. Now, recall that $B = h(A)$ has Property $Q(4, 3)$. Observe that E_2 is homotopic to a part of F, the center of A. We know that F consists of four upper loops U_1, U_2, U_3, U_4, and three lower loops L_1, L_2, L_3. We can label these so that E_2 is homotopic to $E_3 = U_1 \cup U_3 \cup L_1 \cup L_2$. See Figures 12, 13, and 14.

Since $h(A)$ has Property $Q(4, 3)$, it follows that $h(F)$ has Property $P(4, 3)$. Thus, there exist seven points, $p_i \in h(U_i)$, $i = 1,2,3,4$; $q_j \in h(L_j)$, $j = 1,2,3$, such that (1) no two of these points belong to the same loop of $h(F)$ and (2) each p_iq_j intersects both P_1 and P_2. Since E_1 is homotopic to E_2 and E_2 is homotopic to E_3, we have that $h(E_1)$ is homotopic to $h(E_3)$. Consequently, $h(E_1)$ intersects both P_1 and P_2 since $h(A)$ has Property $Q(4, 3)$. But, $h(D_{uv}) \cap P_1 = \varphi$ and $D_{uv} \supset E_1$. This involves a contradiction. Hence, $J_{uv} \cap h^{-1}(P_1) \neq \varphi$. In this way, it follows that $J_{uv} \cap h^{-1}(P_i) \neq \varphi$, $J_{uw} \cap h^{-1}(P_i) \neq \varphi$, and $J_{vw} \cap h^{-1}(P_i) \neq \varphi$ for $i = 1, 2$.

Actually, the argument indicated above shows that (*) <u>if J is the union of two arcs psq and ptq in A and J is homotopic in A to one of J_{uv}, J_{uw}, and J_{vw}, then J intersects both $h^{-1}(P_1)$ and $h^{-1}(P_2)$</u>. We shall use this fact below.

Next, we show that if exactly one are, say xu, of the four arcs xu, xv, yu, and yv of J_{uv} intersects $h^{-1}(P_1)$, then there is a simple closed curve J in A which is homotopic in A to J_{uv} and misses $h^{-1}(P_1)$. We shall use the following theorem due to Bing.

THEOREM (Bing, [4], p. 490). Suppose that M is a 3-manifold with boundary, S_1, S_2 are two disjoint closed subsets of M, and pxq, pyq are homotopic arcs in M such that $pxq \cap S_1 = pyq \cap S_2 = \varphi$. Then there is an arc pzq homotopic to pxq in M such that $pzq \cap (S_1 \cup S_2) = \varphi$.

In this theorem, let pxq be replaced by $xv \cup yv \cup yu = xyu$ while pyq is replaced by xu. And, $S_1 = h^{-1}(P_1) \cap A$, $S_2 = h^{-1}(P_2) \cap A$ and $M = A$. It should be clear that there is a homotopy in A fixed at u which takes xyu onto xu. By use of the above theorem, it follows that there is an arc xzu homotopic to xyu in A such that $xzu \cap$

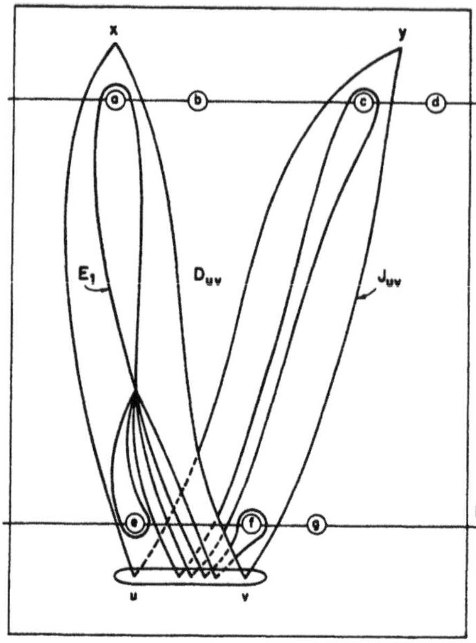

Fig.12

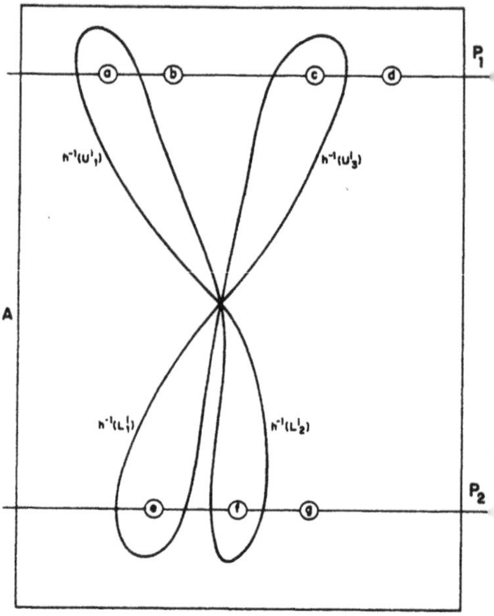

Fig. 13

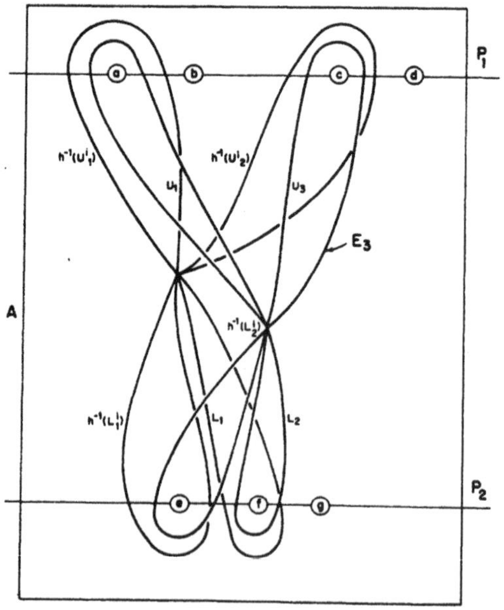

Fig. 14

$[h^{-1}(P_1) \cup h^{-1}(P_2)] = \varphi$. Also, $xzu \cup xyu$ is homotopic in A to J_{uv}. In fact, there is such a homotopy which keeps u fixed. Thus, we have a contradiction to the statement (*) above. In this manner, we conclude that <u>at least two of the four arcs of</u> J_{uv}, J_{uw}, <u>and</u> J_{vw} intersect $h^{-1}(P_1)$ <u>and the other two intersect</u> $h^{-1}(P_2)$.

Our next goal is to show that not both of xu and xv of J_{uv} miss $h^{-1}(P_2)$ while both yu and yv miss $h^{-1}(P_1)$. Suppose that this happens. We apply once again Bing's theorem quoted above. Now, let pxq of this theorem be replaced by $uyv = yu \cup yv$ and pyq be replaced by $uxv = xu \cup xv$. As before, $S_1 = h^{-1}(P_1) \cap A$, $S_2 = h^{-1}(P_2) \cap A$, and $M = A$. Consequently, there is an arc uzv in A homotopic to A to uyv such that $uzv \cap [h^{-1}(P_1) \cup h^{-1}(P_2)] = \varphi$. Also, $J = uzv \cup uxu$ is homotopic in A to J_{uv}. With the statement (*) above, this implies that $J_{uv} \cap h^{-1}(P_2) = \varphi$ which is impossible. We must therefore conclude that the union of two adjacent arcs (of the four) of J_{uv} intersects both $h^{-1}(P_1)$ and $h^{-1}(P_2)$ and that each of the four arcs intersects either $h^{-1}(P_1)$ or $h^{-1}(P_2)$. A similar statement holds true concerning J_{vw} and J_{uw}.

We can color the theta curve T consisting of six arcs with two colors so that each arc has exactly one color and that no two arcs of the same color intersect. Clearly this is impossible. Thus, our assumption that no one of the six arcs of T intersects both $h^{-1}(P_1)$ and $h^{-1}(P_2)$ is false. It follows that the theorem is true. <u>That is, if</u> S <u>is an admissible set of loops, one pair</u> $[h^{-1}(U^i_{m_i}), h^{-1}(L^i_{n_i})]$ <u>has the property that</u> $U^i_{m_i} \cup L^i_{n_i}$ <u>intersects both</u> P_1 <u>and</u> P_2.

Although we shall not present the details, it is not difficult to prove with the use of Theorem 2 that <u>for some</u> m, A_m <u>has either Property</u> $P(2, 3)$ <u>or Property</u> $P(3, 2)$. Is it possible to show that for some m, A_m has Property $P(n, k)$ where $n+k = 6$, $n+k = 7$?

<u>Conjecture</u>. Suppose that h is a homeomorphism of A onto itself that is fixed on $\mathrm{Bd}\, A$. Then there is an element g of the decomposition G such that $h(g)$ intersects both P_1 and P_2.

If the above conjecture is true, then the decomposition space, G, is not homeomorphic to E^3. A proof of this would follow in exactly the same manner as that given for Bing's decomposition space on pages 497-498 of [4]. In fact, Bing actually proves a statement which is more general than his lemma for Theorem 12 on page 497 of [4]. Although the proof is entirely due to Bing, we shall state this theorem and indicate the argument. Some papers have appeared which indicate that the authors are unaware of this fact.

THEOREM (Bing). Suppose that G is an u.s.c. decomposition of E^3 such that the collection H of nondegenerate elements has the properties: (1) each element of H is pointlike (i.e., its complement is homeomorphic

to the complement of a point), (2) in the decomposition space G, H is 0-dimensional, and (3) there is a 3-cell-with-handles such that its interior (bounded component of $E^3 - A$) contains each element of H. If the decomposition space, G, is homeomorphic to E^3, then there is a map f of A onto itself that is fixed at each point of Bd A such that for each point x of A, $f^{-1}(x)$ is an element of G.
A, $f^{-1}(x)$ is an element of G.

Proof: Let h be a homeomorphism of the decomposition space, G, onto E^3. Then there is a map f_1 of E^3 onto itself such that for each point x of E^3, $f_1^{-1}(x)$ is an element of G.

Now, let us suppose that A is as given in the hypothesis. Then there are $n+1$ polyhedral discs $D_1, D_2, \ldots, D_{n+1}$ which are pairwise disjoint such that A is the union of two polyhedral cubes X_1 and X_2 such that $X_1 \cap X_2 = D_1 \cup D_2 \cdots \cup D_{n+1}$. See Figure 15.

Without loss of generality, we suppose that $f_1[\mathrm{Bd}\, A \cup (\bigcup_{i=1}^{n+1} D_i)]$ is locally polyhedral except possibly at points of $f_1(H^*)$ where H^* is the union of the elements of H. The following approximation theorem proved by E. E. Moise allows this.

Theorem 2 of [16]. If K, K' are two triangulated 3-manifolds, U is an open set of K, f is a homeomorphism of U into K', and φ is a continuous positive function defined on U, then there is a locally piecewise linear homeomorphism f' of U into K' such that $\rho(f(x), f'(x)) < \varphi(x)$.

In applying this theorem, we let $K = K' = E^3$, $U = E^3 - H^*$, $\varphi(x) = \rho(x, H^*)/2$, $f = f_1$. The homeomorphism f' on U promised by the theorem may be extended to all of E^3 so as to be a map of E^3 onto itself such that for each point x of E^3, $f'^{-1}(x)$ is an element of G and

$$f'\left[\mathrm{Bd}\, A \cup \left(\bigcup_{i=1}^{n+1} D_i\right)\right]$$

is locally polyhedral except possibly at points of $f'(H^*)$.

From Bing's Theorem 8 of [4], there is a polyhedral disc E_i', $i = 1, 2, \ldots, n+1$, in $f_1(D_i \cup \mathrm{Int}\, A)$ such that $\mathrm{Bd}\, E_i' = \mathrm{Bd}\, f_1(D_1)$. By applying Bing's Theorem 9 of [4] $n+1$ times, we find that there are $n+1$ pairwise disjoint polyhedral discs, $E_1, E_2, \ldots, E_{n+1}$ in f_1

$$f_1\left[A - \mathrm{Bd}\, A \cup \left(\bigcup_{i+1}^{n+1} D_i\right)\right]$$

such that $\mathrm{Bd}\, E_i = f_1(\mathrm{Bd}\, D_i)$. Then

$$\bigcup_{i+1}^{n+1} E_i \cup f_1(\mathrm{Bd}\, A) \text{ is polyhedral.}$$

There is a homeomorphism g of

$$\bigcup_{i=1}^{n+1} E_i \cup f_1(\mathrm{Bd}\ A) \quad \text{onto} \quad \bigcup_{i=1}^{n+1} D_i \cup \mathrm{Bd}\ A$$

that agrees with f_1^{-1} on Bd A. It follows from [1] that g may be extended to a homeomorphism taking the bounded complementary domain U_1 of

$$f_1(\mathrm{Bd}\ A \cap X_1) \cup \left[\bigcup_{i=1}^{n+1} E_i \right]$$

onto $X_1 - \mathrm{Bd}\ X_1$ and the bounded complementary domain U_2 of

$$f_1(\mathrm{Bd}\ A \cap X_2) \cup \left[\bigcup_{i=1}^{n+1} E_i \right]$$

onto $X_2 - \mathrm{Bd}\ X_2$. But,

$$f_1(\mathrm{Bd}\ A) \cup \left[\bigcup_{i=1}^{n+1} E_i \right] \cup U_1 \cup U_2 = f_1(A).$$

Thus, there is a homeomorphism g of $f_1(A)$ onto A that agrees with f_1^{-1} on $f_1(\mathrm{Bd}\ A)$. The map f satisfying the theorem is $g \cdot f_1$.

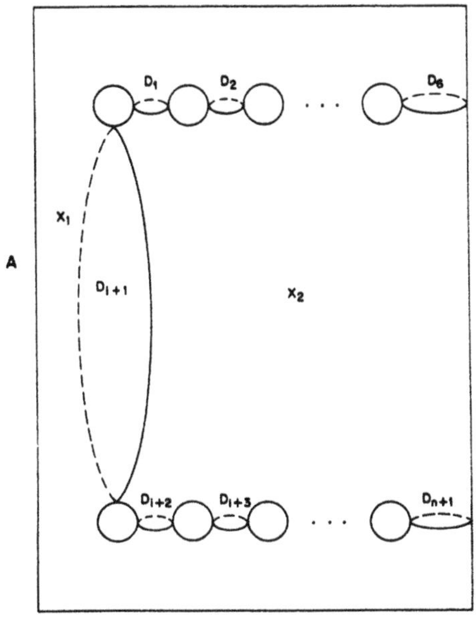

Fig. 15

REFERENCES

[1] J. W. Alexander, "On the subdivision of 3-space by a polyhedron," Proc. Nat. Acad. Sci. USA, vol. 10 (1924), pp. 6-8.

[2] _____, "On the deformation of an n-cell," Proc. Nat. Acad. Sci. USA, vol. 9 (1923), pp. 406-407.

[3] Steve Armentrout, "Upper semicontinuous decompositions of E^3 with at most countably many nondegenerate elements," Annals of Math. vol. 78 (1963), pp. 605-618.

[4] R. H. Bing, "A decomposition of E^3 into points and tame arcs such that the decomposition space is topologically different from E^3," Annals of Math. vol. 65 (1957), pp. 484-500.

[5] _____, "Upper semicontinuous decompositions of E^3, Annals of Math. vol. 65 (1957), pp. 363-374.

[6] _____, "The cartesian product of a certain non-manifold and a line is E^4," Annals of Math. vol. 70 (1959), pp. 399-412.

[7] _____, "Pointlike decompositions of E^3," Fund. Math. vol. 50 (1962), pp. 431-453.

[8] _____, "Decompositions of E^3," Topology of 3-Manifolds and Related Topics, Prentice-Hall (1961), pp. 5-21.

[9] Eldon Dyer and Mary-Elizabeth Hamstrom, "Completely regular mappings," Fund. Math. vol. 45 (1958), pp. 103-118.

[10] M. K. Fort, Jr., "A note concerning a decomposition space defined by Bing, Annals of Math. vol. 65 (1957), pp. 501-504.

[11] David Gillman and Joseph Martin, "Countable decompositions of E^3 into points and pointlike arcs," Notices AMS, vol. 10 (1963), p. 74.

[12] M. E. Hamstrom, "Regular mappings and the spaces of homeomorphisms on a 3-manifold," Memoirs AMS, no. 40 (1961).

[13] L. F. McAuley, "Some upper semicontinuous decompositions of E^3 into E^3," Annals of Math. vol. 73 (1961), pp. 437-457.

[14] _____, "Upper semicontinuous decompositions of E^3 and generalizations to metric spaces," Topology of 3-Manifolds and Related Topics, ed. by Fort, Prentice-Hall (1961), pp. 21-26.

[15] E. E. Moise, "Affine structures in 3-manifolds, IV. Piecewise linear approximations of homeomorphisms," Annals of Math. vol. 55 (1952), pp. 215-222.

[16] _____, "Affine structures in 3-manifolds, V. The triangulation theorem and hauptvermutang," Annals of Math. vol. 56 (1952), pp. 96-114.

[17] R. L. Moore, "Concerning upper semicontinuous collections of continua," *Trans. Amer. Math. Soc.* vol. 27 (1925), pp. 416-428.

[18] C. D. Papakyriakopoulos, "On Dehn's lemma and the asphericity of knots," *Annals of Math.* vol. 66 (1957), pp. 1-26.

[19] Ronald Rosen, "Imbeddings of decompositions of 3-space"(Doctoral Dissertation, University of Wisconsin, August, 1959).

[20] D. E. Sanderson, "Isotopy in 3-manifolds, III. Connectivity of spaces of homeomorphisms," *Proc. Amer. Math. Soc.* vol. 11 (1960), pp. 171-176.

RUTGERS UNIVERSITY

CRUMPLED CUBES

by

L. Lininger

A set C is a crumpled cube iff C is homeomorphic to the union of a 2-sphere and its interior in E^3. If C and D are disjoint crumpled cubes and h is a homeomorphism from Bdy C onto Bdy D, then K is the sum of C and D sewn together along their boundaries iff K is the space obtained from $C \cup D$ by identifying x and $h(x)$ for each x in Bdy C.

An Alexander horned sphere in E^3 with the "hooks" on the inside is an example of a crumpled cube which is not a 3-cell.

The following questions have been raised:

(1) Is the sum of a crumpled cube and a 3-cell S^3?

(2) Under what conditions is the sum of a crumpled cube C and a crumpled cube D sewn together along their boundaries by h topologically S^3?

(3) Under what conditions is the sum of two copies of a crumpled cube C sewn together along their boundaries by the identity homeomorphism topologically S^3?

(4) If the sum of a crumpled cube C and a crumpled cube D sewn together by a homeomorphism h is a 3-manifold, is it topologically S^3?

In 1952, Bing proved that two copies of the Alexander horned sphere sewn together by the identity homeomorphism is topologically S^3 [3]. In 1959, Ball proved that there exists a crumpled cube C and a homeomorphism h (different from the identity) such that, if two copies of C are sewn together by h, the sum is not S^3 [2]. In 1961, Casler showed that if two copies of the Alexander horned sphere are sewn together by any homeomorphism, then the sum is S^3 [8].

In [11] Theorem 1 and Theorem 2 are proved.

THEOREM 1. If a crumpled cube C and a real cube D are sewn together by a homeomorphism h, then the sum is topologically S^3.

This theorem was proved independently by Hosay [10].

THEOREM 2. If C is a crumpled cube, then there exists a pointlike decomposition G of a unit ball B in E^3 such that B/G is homeomorphic to C and each non-degenerate element of G intersects the boundary of B in a one point set.

In fact, it follows that if the crumpled cube C is pointlike in E^3, then there exists a pointlike decomposition G of a tame 3-cell B and pseudo-isotopy h from E^3 onto E^3 that takes B/G onto C, and each non-degenerate element of G is an arc. However, unless some care is taken, $h^{-1}(x)$ will be an arc for each x in Bdy C.

In [12] it is proved that if the sum of two crumpled cubes is a 3-manifold then it is homeomorphic to S^3. The proof given there uses Theorem 1. A somewhat different proof can be obtained from results of Curtis and McMillan, Martin and Armentrout [1], and [9].

An example is known of two copies of the same crumpled cube sewn together with the identity homeomorphism and the result is not S^3, and an example is known of two copies of a crumpled cube C, whose boundary is wild at each point, sewn together by the identity, and the result is S^3 [11].

Frequently it is convenient to know that any crumpled cube in E^3 is the intersection of a sequence $\{M_i\}$ where each M_i is a 3-cell with handles, M_{i+1} is contained in Int M_i, and M_{i+1} is null-homotopic in M_i. See [9].

It is shown in [2] and [12] that every crumpled cube is simply connected. In [12] it is proved that every crumpled cube is 1-ULC and that the sum of two crumpled cubes is connected and simply connected. Example 2 of [11] shows that the interior of a crumpled cube may be simply connected even though the crumpled cube is not a 3-cell. It follows from [4] that if the interior of a crumpled cube is 1-ULC, then the crumpled cube is a 3-cell. It is also known that if the product of a crumpled cube and a 1-cell is a 4-cell, then the crumpled cube is a 3-cell [5].

If a crumpled cube C is a 3-manifold-with-boundary, then by a theorem of M. Brown, C is a 3-cell [6] and [7].

Questions 2 and 3 still are not answered satisfactorily.

(5) If C is a crumpled cube does there exist a pointlike upper semi-continuous decomposition G of a 3-cell B such that

(a) B/G is homeomorphic to C,
(b) each non-degenerate element of G intersects Bdy B in a one-point set, and
(c) the image under the projection map of the non-degenerate elements is a zero-dimensional subset of Bdy C?

(6) Same as Question 5 with the additional conclusion that each non-degenerate element is an arc?

(7) If h is a homeomorphism from Bdy C onto Bdy D and t is a positive number, does there exist a homeomorphism f from Bdy C onto Bdy D such that for each x in Bdy C, $d(f(x), h(x)) < t$ and the sum of C and D sewn together by f is S^3?

(8) Let H denote the set of all homeomorphisms from Bdy C onto Bdy D and let K denote the subset of H such that if f is in K then the sum of C and D sewn together by f is S^3. What can be said about K?

(9) Suppose C is a crumpled cube in E^3. Is $(E^3/C) \times E^1$ topologically E^4? By Theorem 1, it can be assumed that C is a 3-cell in E^3.

(10) Suppose C and D are crumpled cubes sewn together by a homeomorphism h. Is the product of the sum of C and D with E^1 topologically $S^3 \times E^1$?

(11) Suppose C and D are crumpled cubes sewn together by a homeomorphism h. Can the sum of C and D be embedded in E^4?

REFERENCES

[1] S. Armentrout, "Concerning cellular decompositions of 3-manifolds that yield 3-manifolds," to appear in A.M.S. Trans.

[2] B. J. Ball, "The sum of two solid horned spheres," Ann. of Math., vol. 69 (1959) pp. 253-257.

[3] R. H. Bing, "A homeomorphism between the 3-sphere and the sum of two solid horned spheres," Ann. of Math., vol. 56 (1952) pp. 354-362.

[4] ______, "A surface is tame if its complement is 1-ULC," Trans. Amer. Math. Soc., vol. 101 (1961) pp. 294-305.

[5] ______, "A set is a 3-cell if its Cartesian product with an arc is a 4-cell," Proc. Amer. Math. Soc., vol. 12 (1961) pp. 13-19.

[6] M. Brown, "A proof of the generalized Schoenflies theorem," Bull. of the Amer. Math. Soc., vol. 66 (1960) pp. 74-76.

[7] ______, "Locally flat embeddings of topological manifolds," Topology of 3-manifolds and Related Topics, Prentice-Hall, 1961, pp. 83-91.

[8] B. G. Casler, "On the sum of two solid Alexander horned spheres," Abstract 589-15, Notices Amer. Math. Soc., vol. 9 (1962) p. 108.

[9] D. R. McMillan, "A Criterion for Cellularity in a Manifold, Ann. of Math., vol. 79 (1964) pp. 327-337.

[10] N. Hosay, "The sum of a real cube and a crumpled cube is S^3," Abstract 607-17, Notices Amer. Math. Soc., vol. 10 (1963) p. 666.

[11] L. L. Lininger, "Some results on crumpled cubes," Trans. Amer. Math. Soc., vol. 118 (1965) pp. 534-549.

[12] ______, "The sum of two crumpled cubes is S^3 if it is a 3-manifold," to appear.

SEWINGS OF CRUMPLED CUBES WHICH DO NOT YIELD S^3

by

Joseph Martin*

A crumpled cube is a space which is homeomorphic with the closure of the interior of a 2-sphere in S^3. Lininger's paper [5] contains an excellent summary of results which are known about crumpled cubes. The goal of this note is to show that certain sewings of crumpled cubes do not yield S^3. Proofs will appear elsewhere.

Suppose that K is a crumpled cube and $\rho \varepsilon$ Bd K. The statement that ρ is a piercing point of K means that there exists an embedding $f: K \to S^3$ such that f (Bd K) can be pierced by a tame arc at $f(\rho)$. If K is a cell or is the Alexander crumpled cube, then each point of Bd K is a piercing point of K. If K is the Fox-Artin crumpled cube, then K has exactly one non-piercing point.

LEMMA 1: Suppose that K is a crumpled cube and ρ is a piercing point of K. Then if $f: K \to S^3$ is an embedding such that $S^3 - f(K)$ is a cell, f(Bd K) can be pierced by a tame arc at $f(\rho)$.

Let B denote the unit 3-cell in E^3, and if $x \in$ Bd B, let α_x be the straight line interval from the origin to x.

LEMMA 2: Let $h: B \to S^3$ be an embedding. Then h(Bd B) can be pierced by a tame arc at $h(\rho)$ if and only if $h(\alpha_x)$ is tame.

The following lemma is obtained by combining Lemmas 1 and 2

LEMMA 3: Let $h: B \to S^3$ be an embedding. Then $h(\rho)$ is a piercing point of $S^3 - h(B)$ if and only if $h(\alpha_\rho)$ is tame. The main result may now be stated.

THEOREM. Suppose that K and L are crumpled cubes, h: Bd $K \to$ Bd L is a homeomorphism, and the space resulting from identifying K and L

---

* Work on this paper was supported by the National Science Foundation under Grant GP-3857.

by h is S^3. Then if ρ is not a piercing point of K, $h(\rho)$ is a piercing point of L.

In order to establish the theorem, several more lemmas are needed. If X is a subset of E^3, the statement that X is <u>cellular</u> means that there exists a sequence $C_1, C_2, \ldots$ of 3-cells such that $C_{i+1} \subset \text{Int } C_i$, and $X = \cap_1^\infty C_i$.

LEMMA 4: If α is a cellular arc in E^3, then α cannot fail to be locally tame at exactly its two end points.

Lemma 4 was established in collaboration with David Gillman.

If G is an upper semi-continuous decomposition of S^3, we will let S^3/G denote the decomposition space and $P\colon S^3 \to S^3/G$ denote the projection mapping.

LEMMA 5: Suppose that G is an upper semi-continuous decomposition of S^3 into points and compact absolute retracts. Then if υ is a simply connected open set in S^3/G, $P^{-1}(\upsilon)$ is simply connected.

Lemma 5 follows from Smale's version of the Vietoris mapping theorem [7], but a direct elementary proof may be given.

LEMMA 6: Suppose that G is an upper semi-continuous decomposition of S^3 into points and compact absolute retracts, and S^3/G is a 3-manifold. Then each element of G is cellular.

Lemma 6 follows easily from Lemma 5 and McMillan's cellularity criterion, [6]. Lemma 6 may be combined with a theorem of Armentrout [1] to show that if the sum of two crumpled cubes is a 3-manifold, then it is S^3.

It is now an easy matter to establish the theorem. Suppose that K and L are crumpled cubes, $h\colon \text{Bd } K \to \text{Bd } L$ is a homeomorphism, and the resulting identification space is S^3. Now if a 3-annulus is sewn between K and L, it follows from [3] and [4] that the resulting space is S^3. We suppose the annulus is sewn on so that the fibering arcs of the annulus have end points x and $h(x)$. If each of the fibering arcs is shrunk to a point it follows from the hypothesis that the resulting decomposition space is S^3. Now, if ρ is a point of $\text{Bd } K$, it follows from Lemmas 6 and 4 that either the fibering arc from ρ to $h(\rho)$ is locally tame at ρ, or the fibering arc from ρ to $h(\rho)$ is locally tame at $h(\rho)$. It follows from Lemma 3 that either ρ is a piercing point of K or $h(\rho)$ is a piercing point of L. This establishes the theorem.

One would hope that this theorem would describe all of the bad sewings of crumpled cubes. B. J. Ball has an example [2] which shows that this is not the case. Each boundary point of Ball's example is a piercing point, but the sum of this crumpled cube with itself is not S^3.

REFERENCES

[1] S. Armentrout, "Concerning cellular decompositions of 3-manifolds that yield 3-manifolds," to appear in Trans. Amer. Math. Soc.

[2] B. J. Ball, "The sum of two solid horned spheres," Ann. of Math., vol. 69 (1959) pp. 253-257.

[3] N. Hosay, "The sum of a real cube and a crumpled cube is S^3," Abstract 607-17, Notices, Amer. Math. Soc., vol. 10, p.666.

[4] L. Lininger, "Some results on crumpled cubes," Trans. Amer. Math. Soc., vol. 118 (1965)pp. 534-549.

[5] ______, "Crumpled cubes," these notes.

[6] D. R. McMillan, "A criterion for cellularity in a manifold," Ann. of Math., vol. 79 (1964) pp. 327-337.

[7] S. Smale, "A Vietoris mapping theorem for homotopy," Proc. Amer. Math. Soc., vol. 8 (1957) pp. 604-610.

CANONICAL NEIGHBORHOODS IN THREE-MANIFOLDS

by

D. R. McMillan, Jr.*

1. INTRODUCTION

In [6] we gave a necessary and sufficient condition for a compact absolute retract X in the interior of a piecewise-linear (abbreviated: pwl) manifold M^n $(n \geq 5)$ to be cellular in M^n (with respect to piecewise-linear cells). The condition was that for each open set U such that $X \subset U \subset M^n$, there should exist an open set V such that $X \subset V \subset U$ and each loop in $V - X$ is contractible in $U - X$. The same result was obtained for the case $n = 3$, but with the additional restriction that some neighborhood of X be embeddable in E^3.

The motivation for the following theorems is to show that in many useful cases, this disagreeable extra restriction in dimension three can be removed. In particular, if X is a collapsible complex topologically embedded in the interior of a pwl 3-manifold, then X is contained in arbitrarily "small" neighborhoods which are polyhedral cubes-with-handles. This same result is also found to hold for all 2-manifolds-with-non-empty boundary and for some other finite complexes which are not contractible. Perhaps one of the most interesting applications of the above techniques is to topologically embedded closed 2-manifolds, Here we obtain small polyhedral neighborhoods which are homeomorphic to the product of [0. 1] with the given 2-manifold, plus one-dimensional handles.

Section three of this note contains statements of the results obtained thus far. Section four gives a criterion for determining when an open 3-manifold is the union of a monotonically increasing sequence of cubes-with-handles. The relevance of this result to the theorems in Section three is indicated in Section five, and some other questions are raised there also. Complete proofs of the theorems stated here will be given elsewhere, hopefully in a strengthened form.

We wish to thank R. H. Bing for several helpful discussions and for suggesting improved versions of some of the theorems. His approximation theorems (see [1], [2]) are essential in our proofs.

* Research supported in part by grant NSF GP-4125.

2. NOTATION AND DEFINITIONS

The term "manifold" will always be used here in the sense of piecewise-linear (abbreviated: pwl), or combinatorial, manifold. Manifolds are assumed to be connected. A closed manifold is compact and without boundary. An open manifold is non-compact and without boundary. An open manifold M is said to have one end if for each compact set $C \subset M$ there is a compact set D such that $C \subset D \subset M$ and $M - D$ is connected. If M is a pwl manifold with non-empty boundary, a spine of M is any complex onto which M contracts [8].

I will denote the unit interval, and an isotopy will be indicated by h_t $(t \in I)$ in the usual way (i.e., $h_t(x)$ is used in place of $h(x, t)$). If M is a 3-manifold with non-empty boundary, then M is irreducible if each loop in $\mathrm{Bd}\, M$ which is contractible in M is also contractible in $\mathrm{Bd}\, M$. It can be shown that M is reducible (i.e., not irreducible) if and only if there is a polyhedral 2-cell $D \subset M$ such that $D \cap \mathrm{Bd}\, M = \mathrm{Bd}\, D$ and $\mathrm{Bd}\, D$ bounds no 2-cell in $\mathrm{Bd}\, M$. We call such a D a reducing 2-cell for M.

3. PUSHING ARCS AWAY FROM CELLS

If X is a compact, connected subset of the interior of a 3-manifold M, it follows easily from the regular neighborhood theorem [8] of J. H. C Whitehead that X lies in the interior of arbitrarily "small" compact 3-manifolds-with-boundary N. If M is orientable, N may be obtained by attaching disjoint polyhedral 3-cells $B_1, B_2, \ldots, B_k$ to a cube-with-handles H in such a way that each $B_i \cap H = (\mathrm{Bd}\, B_i) \cap (\mathrm{Bd}\, H)$ is an annulus C_i (since N is equivalent to the regular neighborhood in M of some 2-complex). If we bore holes in N along the "core" of each B_i from one component of $(\mathrm{Bd}\, B_i) - C_i$ to the other, what we have left is equivalent to H. Formally:

LEMMA 1. Let N be a compact, orientable 3-manifold with non-empty boundary. Then N contains a finite, disjoint collection of polygonal arcs $A_1, A_2, \ldots, A_k$, such that $A_i \cap \mathrm{Bd}\, N = \mathrm{Bd}\, A_i$, for each i, and the closure of N minus a thin tubular neighborhood of each A_i is a cube-with-handles.

Hence, one way of obtaining a cube-with-handles containing X in its interior is to show the existence of a pwl homeomorphism $h\colon N \to N$ such that $h|\mathrm{Bd}\, N$ is the identity and $X \cap h(A_i) = \emptyset$, for each i. For then the above holes may be bored without hitting X. The problem of producing such an h reduces to the following:

QUESTION. If N is a compact, orientable piecewise-linear 3-manifold-with-boundary, X is a compact subset of $\mathrm{Int}\, N$, A is a polygonal arc in N such that $A \cap \mathrm{Bd}\, N = \mathrm{Bd}\, A$, and ε is a positive number, when does there exist a piecewise-linear homeomorphism h of N onto N such that h is the identity outside an ε-neighborhood of X and $X \cap h(A) = \emptyset$?

If X is a 1-dimensional compact absolute neighborhood retract or a 0-dimensional compact set, we can find such an h that moves each point of N less than ε (see the proof of [3: Lemma 3] in the case $n = 3$). If X is a 2-cell we can "pipe" A around the boundary of X without moving points of N far from X. This piping is done along tame arcs in X running out to Bd X (the existence of such arcs is guaranteed by Martin [4]). Using such techniques as these (the case when X is a 3-cell is the most difficult), we obtain:

THEOREM 1. Let X be an i-cell ($i = 1, 2,$ or 3) topologically embedded in the interior of a piecewise-linear 3-manifold M, and let $D \subset$ Bd X be an $(i-1)$-cell. Let $A_1, A_2, \ldots, A_k$ be a finite disjoint collection of polygonal arcs in $M - D$ with each Bd $A_j \subset M - X$. Then there is a compact set $C \subset X - D$ such that, for given $\varepsilon > 0$, there exists a piecewise-linear homeomorphism h of M onto M such that each $h(A_j) \subset M - X$ and h is the identity outside an ε-neighborhood of C.

Let us say that $X \subset \text{Int } M$ has the <u>cube-with-handles property</u> if $X = \bigcap_{i=1}^{\infty} H_i$, where $H_i \subset \text{Int } H_{i-1} \subset M$ and H_i is a polyhedral cube-with-handles. Then it follows easily from Theorem 1 that:

THEOREM 2. Let K be a finite simplicial complex, L a subcomplex of K such that K contracts to L, and $g\colon K \to \text{Int } M$ a homeomorphism, where M is a piecewise-linear 3-manifold. Then, if $g(L)$ has the cube-with-handles property, so does $g(K)$.

An immediate corollary of Theorem 2 is that, in any orientable 3-manifold, topologically embedded collapsible complexes, 2-manifolds with non-empty boundary, and (fortunately) cubes-with-handles, always have the cube-with-handles property. From Theorem 1 we also obtain:

THEOREM 3. Let M be an orientable, piecewise-linear 3-manifold and suppose $X \subset \text{Int } M$ is a topologically embedded, 2-sided, closed 2-manifold. Then, for given $\varepsilon > 0$, the ε-neighborhood of X contains a polyhedral 3-manifold H such that $X \subset \text{Int } H$ and H is equivalent to $X \times I$ plus 1-dimensional handles of diameter less than ε.

By saying that the handles of H are of diameter less than ε, we mean that there is a collection $K_1, K_2, \ldots, K_p$ of mutually exclusive polyhedral cubes-with-handles in H such that each K_i has diameter less than ε, and if we let N denote the closure of $H - \bigcup K_i$, then N is equivalent to $X \times I$, and $K_i \cap N$ is a 2-cell in $(\text{Bd } K_i) \cap (\text{Bd } N)$ for each i.

A similar result can be obtained when X is a compact 3-manifold with non-empty boundary, topologically embedded in Int M. For example, if

X is a topological 3-cell in Int M, then the ε-neighborhood of X contains a polyhedral cube-with-handles H such that $X \subset$ Int H and H has handles of diameter less than ε. The same result holds, in fact, even if X is a crumpled cube, i.e., the closure of a complementary domain of a 2-sphere in S^3.

4. A THEOREM ON OPEN MANIFOLDS

The following theorem and its proof are a generalization of [5; Theorem 1] (see also [6; Lemma 1]).

THEOREM 4. Let M be an orientable open piecewise-linear 3-manifold with one end and such that each polyhedral 2-sphere in M bounds a 3-cell in M. Then, a necessary and sufficient condition that M can be expressed as the union of a sequence of polyhedral cubes-with-handles H_1, H_2, ..., where $H_i \subset$ Int H_{i+1}, is that the following condition hold:

For each compact set $C \subset M$, there is a compact set D such that $C \subset D \subset M$ and each polygonal arc in M having endpoints in $M - D$ is homotopic in M (with endpoints fixed) to an arc in $M - C$.

Proof: To see the necessity, let C be given as above and choose for D any polyhedral cube-with-handles such that $C \subset$ Int D. Now, if a polygonal arc A with endpoints in $M - D$ is given, we choose a 1-dimensional spine G for D in (Int D) - A. Then there is an isotopy $h_t\colon M \to M$ such that h_0 is the identity, h_t is the identity outside a small neighborhood of D not containing Bd A, and $h_1(A) \subset M - D$. This isotopy may be pictured as swelling a small tubular neighborhood of G until this neighborhood coincides with D. Thus we have shown more than is required to prove the condition necessary.

To show the condition sufficient, it is enough to prove that if Σ is a finite, disjoint collection of compact polyhedral 3-manifolds-with-boundary in M, then Σ^*, the union of the elements of Σ, lies in a cube-with-handles in M. Since a non-orientable surface (i.e., closed 2-manifold) in an orientable 3-manifold cannot be two-sided, the boundary of each element of Σ is an orientable surface. Define the non-negative integer $c(\Sigma)$ to be $\Sigma_{n>0}\, n^2 g(n)$, where $g(n)$ is the number of surfaces of genus n in the collection of boundary components of elements of Σ. We shall use induction on $c(\Sigma)$. Since each surface in M separates M into exactly two components and the closure of exactly one of these is compact, we may assume that each element of Σ has connected boundary.

If $c(\Sigma) = 0$, then each element of Σ is a polyhedral 3-cell, and each of these can be expanded slightly and then joined by tubes to give a polyhedral 3-cell containing Σ^*. Suppose then that $c(\Sigma) > 0$, but that if Σ_0 is any finite, disjoint collection of compact, polyhedral 3-manifolds-with-boundary in M with $c(\Sigma_0) < c(\Sigma)$ then $\Sigma_0{}^*$ lies in a polyhedral cube-with-handles in M.

If the closure of $M - \Sigma^*$ is a reducible 3-manifold, we can thicken nicely one of its reducing 2-cells and add the resulting 3-cell to the appropriate element of Σ to obtain a collection Σ_0 with $c(\Sigma_0) < c(\Sigma)$ and $\Sigma^* \subset \Sigma_0{}^*$. Hence, in this case we are done.

If some element R of Σ is reducible, we cut R along one of its reducing 2-cells E to obtain a collection Σ_0 with $c(\Sigma_0) < c(\Sigma)$. By induction, there is a cube-with-handles H such that $\Sigma_0{}^* \subset \operatorname{Int} H$. We then can adjust H near the cut so that $H \cup \Sigma^*$ is a polyhedral cube-with-handles. This adjustment is done with a pwl homeomorphism h of M onto M which is the identity outside a small neighborhood of E. We describe h as follows. Choose a spine for R of the form $R_0 \cup A$, where R_0 is R cut along E and A is a polygonal arc in general position relative to H and with $A \cap R_0 = \operatorname{Bd} A$. Let N be a nice thin tube around A, so that $R_0 \cup N$ is a spine for R and $H \cup N$ is a cube-with-handles. The homeomorphism h then expands $R_0 \cup N$ until it coincides with R. Hence, the proof can be completed in this case also.

If each element of Σ is a cube-with-handles, we can obtain a cube-with-handles containing Σ^* by joining these manifolds with tubes as before. If some element K of Σ is not a cube-with-handles, we show that one of the preceding two cases must occur. Let D be the compact set promised by our hypothesis for the compact set Σ^*. Since K is not a cube-with-handles, there is, by [7; Theorem 19.1], a polygonal arc B in K such that $B \cap \operatorname{Bd} K = \operatorname{Bd} B$ and B is <u>not</u> homotopic in K (with endpoints fixed) to any arc in Bd K (our hypotheses make it unnecessary to assume the truth of the Poincaré conjecture in Papakyriakopoulos' theorem). Extend B to a polygonal arc J such that $J \cap \Sigma^* = B$ and $\operatorname{Bd} J \subset M - D$.

By hypothesis, J is homotopic in M (with end-points fixed) to an arc in $M - \Sigma^*$. Hence, if we use the notations

$$X = \{(x, y) \in I \times I \mid y = 0\} ,$$

$$Y = \{(x, y) \in I \times I \mid x = 0 \text{ or } 1\}, \text{ and}$$

$$Z = \{(x, y) \in I \times I \mid y = 1\} ,$$

there is a pwl mapping $f\colon I \times I \to M$ such that $f| X \cup Y$ is a homeomorphism onto J, with $f(X) = B$, and such that $f(Z) \subset M - \Sigma^*$. Further, since J could be taken in general position relative to the boundary surfaces of Σ^*, we may assume that the inverse image under f of all the boundary surfaces of Σ^*, consists of a finite disjoint collection of polygonal simple closed curves in $\operatorname{Int}(I \times I)$ plus a polygonal arc missing these curves and meeting $\operatorname{Bd}(I \times I)$ precisely in $(0, 0)$ and $(1, 0)$.

If neither of the two cases considered previously occurs, we find from the definition of irreducibility that f restricted to an "inner" one of the above simple closed curves is a loop in the boundary of an element of Σ which is contractible in that boundary surface. Such a curve can be eliminated by the standard procedure of exchanging singular discs. Continuing, we could eliminate all the simple closed curves in the inverse image under f of the boundary surfaces of Σ^*, and find that B is homotopic in K to an arc in Bd K. This contradiction completes the proof.

5. QUESTIONS

1. Think of a better name for the "cube-with-handles property."

2. Is the cube-with-handles property a topological property? That is, if $X \subset \text{Int } M$ has the cube-with-handles property and $h\colon X \to \text{Int } N$ is an embedding, does $h(X)$ have the cube-with-handles property?

3. A special case of question two: If each of $h_i\colon K \to \text{Int } M_i$ $(i = 1, 2)$ is a pwl embedding of the finite complex K and the regular neighborhood of $h_1(K)$ in M_1 is a 3-cell, is the regular neighborhood of $h_2(K)$ in M_2 a 3-cell?

4. Another special case of question two: It is known that if the "dunce's hat" (a contractible, but not collapsible, 2-complex) is piecewise-linearly embedded in any 3-manifold, then its regular neighborhood is a 3-cell. Does the dunce's hat have the cube-with-handles property?

5. If K is a finite complex and $f\colon K \to \text{Int } M$ is a topological embedding, does there exist $\varepsilon > 0$ such that for any two pwl ε-approximations $g_i\colon K \to \text{Int } M$ $(i = 1, 2)$ to f, the regular neighborhoods of $g_1(K)$ and $g_2(K)$ in M are equivalent? If so, does $f(K)$ have arbitrarily small polyhedral neighborhoods equivalent to this (unique) regular neighborhood plus small 1-dimensional handles?

6. Theorem 4 can sometimes be used to express $X \subset S^3$ as the intersection of "unknotted" (i.e., canonically embedded) cubes-with-handles. To do this, it suffices to show that $S^3 - X$ is the union of a monotonically increasing sequence of cubes-with-handles. Can Theorem 4 be used to find some useful class of X's for which this can be done? For example, what if X is a topological cell? The only obvious application is when $X \subset S^3$ is a compact absolute retract and $S^3 - X$ is simply connected.

REFERENCES

[1] R. H. Bing, "Approximating surfaces from the side," Ann. of Math., 77 (1963) pp. 145-192.

[2] ______, "Each disc in E^3 contains a tame arc," Amer. J. Math., 84 (1962) pp. 583-590.

[3] M. L. Curtis and D. R. McMillan, "Cellularity of sets in products," Mich. Math. Jour. 9 (1962) pp. 299-302.

[4] Joseph Martin, "Tame arcs on discs," Proc. Amer. Math. Soc., 16 (1965) pp. 131-133.

[5] D. R. McMillan, Jr., "Cartesian products of contractible open manifolds," Bull. Amer. Math. Soc., 67 (1961) pp. 510-514.

[6] ______, "A criterion for cellularity in a manifold," Ann. of Math., 79 (1964) pp. 327-337.

[7] C. D. Papakyriakopoulos, "On solid tori," Proc. London Math. Soc., (3) 7 (1957) pp. 281-299.

[8] J. H. C. Whitehead, "Simplicial spaces, nuclei, and m-groups," Proc. London Math. Soc., 45 (1939) pp. 243-327.

BOUNDARY LINKS

by

N. Smythe

I. In 1936 Eilenberg [1] introduced the concepts of weak linking and n-linking. In summary

a) A polyhedral link L of two components ℓ_1, ℓ_2 in S^3 is said to be 1-linked if for every pair of polyhedra P_1, P_2 such that $\ell_i \sim 0$ in P_i, we have $P_1 \cap P_2 \neq \emptyset$. (0-linked means the components have non-zero linking number. In general, ℓ_1 is n-linked with ℓ_2 if every polyhedron P such that $\ell_1 \sim 0$ in P, contains a cycle which is (n-1)-linked with ℓ_2.) The link is 1-linked but not 0-linked.

b) L is weakly linked (faiblement enlace) if every pair of 1-cycles γ_1 in V_1, γ_2 in V_2, where V_i is a regular neighborhood of ℓ_i, such that $\gamma_i \not\sim 0$ in V_i, are 1-linked.

Using this theory of multicoherence, Eilenberg was able to characterize weakly linked links as: L is weakly linked if and only if there exists no homomorphism of $\pi_1(S^3 - L)$ onto a free group of rank two.

The question was raised, does there exist a link which is 1-linked but not weakly linked?

II. Some time ago, in an unpublished manuscript, R. H. Fox considered a generalization of 1-linking to links of more than two components.

A <u>boundary link</u> $L = \ell_1 \cup \ell_2 \cup \cdots \cup \ell_n$ in S^3 is one whose components bound disjoint orientable surfaces.

Clearly, a link of 2 components is a boundary link if and only if it is <u>not</u> 1-linked.

These links have various nice properties; e.g., the elementary ideals $\mathcal{E}_0(t_1 \cdots t_n), \ldots \mathcal{E}_{n-1}(t_1 \cdots t_n)$ are all 0, and $\mathcal{E}_n(t, t, \ldots, t)$ is principal, generated by a knot polynomial. Also the longitudes of L lie in the second commutator subgroup of $\pi_1(S^3 - L)$.

III. Weak linking can also be generalized to links of many components, although the concept becomes somewhat unrecognizable. [3] $L = \ell_1 \cup \ell_2 \cup \ldots \cup \ell_n$ in S^3 (with solid torus regular neighborhoods $V_1, V_2, \ldots, V_n$) will be called an <u>homology boundary link</u> if

1) There exist disjoint orientable surfaces $S_1 \cdots S_n$ in $S^3 - \cup \operatorname{Int} V_i$ such that $S_i \cap V_j = \dot{S}_i \cap \dot{V}_j$ consists of a number of simple closed curves $\lambda_{ij1}, \lambda_{ij2}, \ldots, \lambda_{ijm}$ (with orientation induced from S_i) which are longitudes of V_j.

2) $\dot{S}_i = \cup_{j,k} \lambda_{ijk}$

3) $\Sigma_k \lambda_{ijk} \sim 0$ in V_j if $j \neq i$

4) $\Sigma_k \lambda_{ijk} \sim \ell_i$ in V_i.

(Shrinking V_i down to its core ℓ_i, we see that L is a boundary link except that the surfaces are allowed to have singularities along the components of L.)

Then a link of 2 components is an homology boundary link if and only if it is not weakly linked. This follows from

THEOREM: L is a homology boundary link if and only if there exists a homomorphism $f: \pi_1(S^3 - L) \to F(n)$ onto a free group of rank n. Furthermore, L is a boundary link if and only if there exists meridians $\alpha_1, \ldots, \alpha_n$ of $\ell_1, \ldots, \ell_n$ such that $f(\alpha_1), \ldots, f(\alpha_n)$ freely generate $F(n)$.

To construct such a homomorphism given a homology boundary link, first thicken the surfaces S_i to open sets $N_i = \mathring{S}_i \times (0, 1)$; then $S^3 - \cup V_i$ can be mapped onto a wedge of n circles by mapping points outside $\cup N_i$ to the vertex, and a point of $S_i \times \{t\}$ to the point t of the i-th circle. Since $\cup S_i$ cannot separate $S^3 - \cup V_i$, the induced mapping of fundamental groups is the required homomorphism.

The idea of the proof in the other direction is to construct a retraction of $S^3 - \cup V_i$ onto a wedge of n circles. The surfaces will then appear as the inverse images of non-critical points (non-vertices). In the case of a boundary link, the n-leafed rose must be embedded in $S^3 - L$ as a set of meridians.

This algebraic characterization leads immediately to the fact that a homology boundary link has an Alexander matrix $A(t_1, \ldots, t_n)$ with n columns of zeros (if the link is actually a boundary link, these columns correspond to meridians of $\pi_1(S^3 - L)$). Thus again $\mathcal{E}_0 = \mathcal{E}_1 = \cdots = \mathcal{E}_{n-1} = 0$.

Also, the homomorphism f induces isomorphisms $f_\alpha: G/G_\alpha \approx F/F_\alpha$ of the quotient groups of the lower central series, where $G = \pi_1(S^3 - L)$. Since $F_\omega = 1$, we have $f_\omega: G/G_\omega \approx F$, and $G_\omega = \ker f$. The longitudes of L lie in G_ω, since they lie on the surfaces. The Milnor isotopy invariants $\bar{\mu}(i_1 \cdots i_k)$ of L are therefore all 0. (In fact much more can be said: it turns out that any link which is isotopic to a homology boundary link under a "nice" (e.g., polyhedral) isotopy, is an homology boundary link.) See [3], [4].

Eilenberg's question becomes: Does there exist a homology boundary link which is not a boundary link?

The answer is yes. The link of the figure is not a boundary link (it is easy to show that one of its longitudes does not lie in G''). But

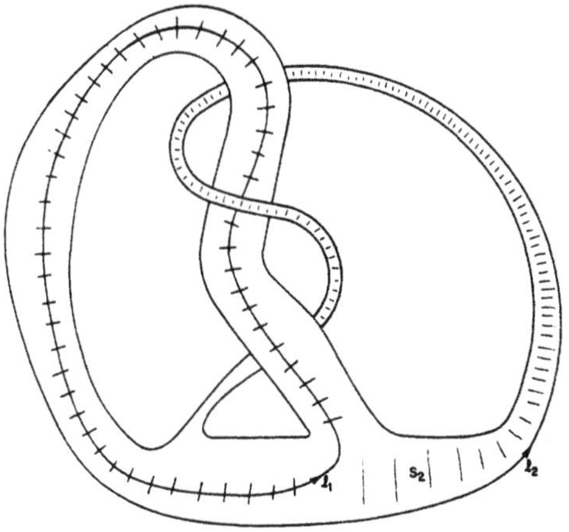

it is a homology boundary link (either by constructing the surfaces, or constructing a map of G onto a free group of rank 2—but of course there is no such map under which a pair of meridians are free generators).

The surface S_2 is shown in the diagram, to be thought of with two boundaries lying parallel to ℓ_1 on a regular neighborhood V_1 of ℓ_1. The surface S_1 is to have 3 boundaries, all longitudes of V_1, one having opposite orientation to the others (two lie "above" S_2, and one "below").

<u>Questions</u>:

(1) (Fox) Suppose the longitudes of L lie in G''. Is L a boundary link?

(2) Similarly, suppose $A(t_1 \cdots t_n)$ has n columns of zeros. Is L a homology boundary link (or a boundary link, if the columns correspond to meridians)?

(3) Is every link isotopic to a boundary link also a boundary link (under a "nice" isotopy)?

(4) Is there a corresponding theory for links in higher dimensions?

(5) Generalize n-linking to arbitrary numbers of components and characterize algebraically.

(6) There are a number of questions related to n-linking—see Fox [2].

(7) (Milnor) What does G_ω look like? Can invariants (say isotopy invariants) of G_ω be found to distinguish between homology boundary links?

REFERENCES

[1] S. Eilenberg, "Multicoherence I, II," Fundamenta Math., 27 (1936) p. 153; 29 (1937) p. 107.

[2] R. H. Fox, Some Problems in Knot Theory, Topology of 3-Manifolds, Prentice-Hall, 1962.

[3] N. Smythe, Isotopy Invariants of Links, Ch. IV, Princeton Ph.D. Thesis (1965).

[4] J. Milnor, "Isotopy of Links, Algebraic Geometry and Topology" (Lefschetz Symposium) Princeton University Press, Princeton, 1957.

SURFACES IN E^3*

by

C. E. Burgess

We present a brief summary, with references, of developments during the last fifteen years that are related to tame embeddings of 2-manifolds in 3-manifolds. This is the basis of a twenty minute talk presented in the Special Session on Recent Developments in Topology at the meeting of the American Mathematical Society at Iowa City on November 27, 1965. For simplicity in the presentation, we restrict our discussion to 2-spheres in E^3, but most of the theorems can be extended to 2-manifolds in a 3-manifold.

We have included an extensive list of references on this topic, but we presume that there are some serious omissions and some lack of current information in this list. We find it necessary to place some restriction on the number of papers listed here, and we suggest that other work essential to this development can be found in the references cited in the papers that are included in this list. For example, see Harrold's expository paper [37].

1. Fundamental work related to tame imbeddings of 2-spheres in E^3. A 2-sphere S is defined to be tame in E^3 if there is a homeomorphism h of E^3 onto itself that carries S onto the surface of a tetrahedron. Alexander [1] showed in 1924 that every polyhedral 2-sphere in E^3 is tame under this definition. Graeub [30] and Moise [41, II], working independently, showed in 1950 that every polyhedral sphere in E^3 can be carried onto the surface of a tetrahedron with a piecewise linear homeomorphism of E^3 onto itself.

A fundamental characterization of tame spheres in E^3 was given by Bing in 1959 with the following theorem. (In the various theorems on characterizations of tame surfaces in E^3, we do not include the obvious statement that every tame sphere satisfies the requirements of the characterization.)

1.1 A 2-sphere is tame if it can be homeomorphically approximated in each of its complementary domains. (Bing [10].)

The proof of this theorem depended upon Bing's Approximation Theorem for Spheres [8], which was later generalized to the following Side Approximation Theorem.

---

* This paper was not presented at the seminar, but it is included by invitation since it treats recent developments and questions discussed at the seminar. This work was supported by NSF Grant GP-3882.

1.2 If S is a 2-sphere in E^3 and $\varepsilon > 0$, then there exist (1) a polyhedral 2-sphere S', (2) a finite collection $D_1, D_2, \ldots, D_n$ of disjoint ε-discs on S', and (3) a finite collection of disjoint ε-discs $E_1, E_2, \ldots, E_n$ on S such that

(1) S' is homeomorphically with ε of S,

(2) $S' - \Sigma D_i \subset \text{Int } S$, and

(3) $S - \Sigma E_i \subset \text{Ext } S'$.

Furthermore, Int S and Ext S' can be replaced with Ext S and Int S', respectively, in conclusions (2) and (3) above. (Bing [14], [17].)

Essential to some of the recent developments on tame surfaces are papers on locally tame sets [7], [41, VIII], triangulation of 3-manifolds [9], [41, V], and Dehn's Lemma and the Sphere Theorem [42], [43]. Brown's work on the generalized Shoenflies Theorem [18] and on locally flat imbeddings [19] is applicable, but for surfaces in E^3, or in 3-manifolds comparable results can be obtained from the work of Bing and Moise cited above.

2. Examples of wild 2-spheres in E^3. A 2-sphere in E^3 is defined to be wild if it is not tame. Early examples of wild spheres in E^3 were given by Antoine [4] in 1921 and by Alexander [2] in 1924. The set of wild points in Antoine's example is a wild Cantor set, and in Alexander's example this set is a tame Cantor set. A third type of wild sphere, which has only one wild point, was described by Fox and Artin [27] in 1948. In 1961, Bing [12] described a wild 2-sphere in which every arc is tame. Examples of other types of wild spheres can be found in papers by Alford [3], Ball [5], Bing [6], [16], Cannon [23], Casler [24], Fort [26], Gillman [28], and Martin [40]. Of course other wild spheres can be described by combining some of the methods used in describing these examples.

Relatively little has been done on the equivalence of imbeddings of wild spheres in E^3. For horned spheres, work by Ball [5] and Casler [24] shows that the horned sphere described by Ball [5] is imbedded differently from the one described by Alexander [2], and Cannon [23] has describd a 3-horned sphere that is imbedded differently from either of these. These three examples are free of any knotting in the horns.

3. Other characterizations of tame spheres in E^3. Let S be a 2-sphere in E^3. Then S is tame if the requirements of any one of the following statements are fulfilled.

3.1 Every disc in S has both the strong enclosure property and the hereditary disc property. (Griffith [31].)

3.2 S is locally peripherally unknotted; i.e., for each $p \in S$ and each $\varepsilon > 0$, there exists a tame 2-sphere S' such that $p \in \text{Int } S'$, diam $S' < \varepsilon$, and $S \cap S'$ is a simple closed curve. (Harrold, [32], [33].)

3.3 The complement of S is 1-ULC. (Bing [11].)

3.4 S can be deformed into each complementary domain; i.e., for each component U of E^3-S, there is a homotopy h: $S \times I \to cl(U)$ such that $h(x, 0) = x$ and $h(x, t) \in U$ for $0 < t \leq 1$. (Hempel [35].)

3.5 S is the image of a tame sphere S' under a map f of E^3 onto itself such that $f|S'$ is a homeomorphism and $f(E^3 - S') = E^3 - S$. (Hempel [35].)

3.6 S can be locally spanned from each complementary domain; i.e., for each $p \in S$, for each $\varepsilon > 0$, and for each component U of E^3- S, there exist discs D and R such that $p \in \text{Int } R \subset S$, $\text{Int } D \subset U$, $D \cap R = \text{Bd } D = \text{Bd } R$, and $\text{diam } (D + R) < \varepsilon$. (Burgess [21].)

3.7 S can be uniformly locally spanned in each complementary domain; i.e., for each component U of E^3 - S and for every $\varepsilon > 0$ there exists a $\delta > 0$ such that for any δ-disc R on S and for any $\alpha > 0$, there is an ε-disc D in U such that Bd R can be shrunk to a point in D + (some α-neighborhood of Bd R). (Burgess [21].)

3.8 (Let K be a Sierpinski Curve; i.e., K is a universal one-dimensional plane curve.) For each component U of E^3- S and each $p \in S$, there is a homeomorphism h: $K \times I \to E^3$ such that $h(K \times 0) \subset S$, p is an inaccessible point (relative to S) of $h(K \times 0)$, and $h(K \times t) \subset U$ for $0 < t \leq 1$. (Burgess [21].)

3.9 S can be locally spanned in each complementary domain on tame simple closed curves; i.e., for each component U of E^3 - S, for each $p \in S$, and for each $\varepsilon > 0$, there exists an ε-disc R in S such that (1) $p \in \text{Int } R \subset S$, (2) Bd R is tame, and (3) for each $\alpha > 0$ there is an ε-disc D in U such that Bd R can be shrunk to a point in D + (an α-neighborhood of Bd R). (Loveland [38].)

3.10 For each $p \in S$ and each $\varepsilon > 0$ there is a 2-sphere S' satisfying the following requirement: (1) $p \in \text{Int } S'$, (2) diam S' $< \varepsilon$, (3) $S \cap S'$ is a continuum, and (4) S can be side-approximated missing $S \cap S'$. (Loveland [38].) (See the next section for a definition of the requirement in (4).)

4. Tame subsets of 2-spheres in E^3. A subset H of a 2-sphere S in E^3 is defined to be tame if H is a subset of a tame sphere. We say that S can be side-approximated missing H if, in the Side Approximation Theorem stated above, it is further required that $H \subset S - \Sigma E_1$. Bing's methods of using the Side Approximation Theorem to prove the following theorem have been essential in subsequent work on tame subsets of spheres.

4.1 If S is a 2-sphere in E^3 and $\varepsilon > 0$, then there exists a tame Sierpiński curve K in S such that each component of $S - K$ has a diameter less than ε. (Bing [13].)

The following four theorems have been announced in the chronological order in which they are listed, and each of the last three is a generalization of the preceding one.

4.2 If the 2-sphere S in E^3 can be side-approximated missing the arc H in S, then H is tame. (Gillman [29].)

4.3 If S is a 2-sphere in E^3, H is a closed subset of S such that the set of all diameters of the components of H has a positive lower bound, and S can be side-approximated missing H, then H is tame. (Hosay [37].)

4.4 If the 2-sphere S in E^3 can be side-approximated missing the closed subset H of S and $\varepsilon > 0$, then there exists a tame subcontinuum M of S and a null sequence D_i of ε-discs in S such that (1) $M = S - \Sigma \text{ Int } D_i$ and (2) $H \subset M - \Sigma D_i = S - \Sigma D_i$. (Loveland [39].)

4.5 If each of $H_1, H_2, \ldots, H_n$ is a closed subset of the 2-sphere S in E^3 and, for each i, S can be side-approximated missing H_i, then ΣH_i is tame and S may be side-approximated missing ΣH_i. (Loveland [39].)

Proofs of the following two theorems depend upon the methods used in proving the above four theorems.

4.6 If S is a 2-sphere in E^3, H is a closed subset of S such that the set of all diameters of the components of H has a positive lower bound, and for each $\varepsilon > 0$ there is a $\delta > 0$ such that each δ-simple closed curve in $E^3 - S$ can be shrunk to a point in an ε-subset of $E^3 - H$, then H is tame. (Hosay [37] and Loveland [39].)

4.7 If S is a 2-sphere in E^3, H is a closed subset of S such that the set of all diameters of the components of H has a positive lower bound, and S can be locally spanned at each point of H from each complementary domain of S, then H is tame. (Loveland [38].)

5. Spheres in E^3 which are tame except on a tame set. Let S be a 2-sphere in E^3 and let H be a tame subset of S such that S is locally tame at each point of $S - H$. We present here some additional requirements on H which imply that S is tame. The foundation for the first result of this type can be found in Moise's work [41, VIII], and this led to the following generalization.

5.1 S is tame if H is a tame finite graph. (Doyle and Hocking [25].)

Using consequences of Dehn's Lemma, the Side Approximation Theorem, and his earlier theorem characterizing tame spheres, Bing proved the following more general theorems.

5.2 S Is tame if H is a tame Sierpinski curve. (Bing [15].)

5.3 S is tame if H is the union of a finite number of sets each of which is either a tame finite graph or a tame Sierpinski curve. (Bing [15].)

Other results have been obtained where H is the intersection of a tame sphere with S.

5.4 If S is a 2-sphere in E^3 and S' is a tame 2-sphere such that (1) S is locally tame at each point of S - S', (2) $S \cap S'$ does not separate S, and (3) the set of all diameters of components of $S \cap S'$ has a positive lower bound, then S is tame. (Hosay [36].)

5.5 If C is a 3-cell in E^3 and S is a 2-sphere in C such that S is locally tame from Int S at each point of S-Bd C, then S + Int S is a 3-cell. (Burgess [20], [22].)

The interior of a disc D in E^3 is defined to be <u>locally tame from one side</u> if for each $p \in$ Int D there exists a disc D' and a 3-cell E such that $p \in$ Int $D' \subset D$ and $E \cap D = D' \subset$ Bd E.

5.6 If C is a 3-cell in E^3 and D is a disc in C such that D is locally tame at each point of D - Bd C, then Int D is locally tame from one side. (Burgess [22].)

5.7 If p is an isolated wild point of a 2-sphere S in E^3, then S is locally tame at p from one of the components of E^3- S. (Harrold and Moise [34].)

6. <u>Questions on tame spheres and on tame subsets of spheres in E^3.</u> The questions raised here have arisen in papers, seminars, and informal discussion in various places. We include a reference where we know a question has appeared in the literature.[1]

Which of the following conditions imply that a 2-sphere S in E^3 is tame?

6.1 S can be pierced by a tame disc at each arc in S.

6.2 S can be pierced by a tame annulus at each simple closed curve in S.

6.3 S can be approximated with a map of S in each complementary domain of S [35, p. 280].

6.4 Each Sierpinski curve in S can be pushed with a homotopy (an isotopy) into each component of E^3- S.

---

1 Some of these questions were included in an address by R. H. Bing at the International Congress of Mathematicians in 1962. (See pp. 457-548 of the Proceedings of that Congress.)

6.5 For each $p \in S$, for each component U of $E^3 - S$, and for each $\varepsilon > 0$, there is an ε-disc R such that $p \in \text{Int } R \subset S$ and the identity map on Bd R can be shrunk to a point in an ε-subset of Bd $R + U$ [21, p. 89].

6.6 S is locally tame at each point of $S - H$, where H is a non-degenerate tame subcontinuum of S.

6.7 S is locally tame at each point of $S - H$, where H is a tame closed subset of S having no degenerate component.

6.8 S can be locally spanned in each component of $E^3 - S$.

6.9 For each $p \in S$ and each $\varepsilon > 0$, there exists a tame 2-sphere S' such that $p \in \text{Int } S'$, diam $S' < \varepsilon$, and $S \cap S'$ is a continuum.

6.10 For each $p, q \in S$, there is a homeomorphism of E^3 onto itself that carries S onto itself and p onto q [28, p. 253].

6.11 Each horizontal plane that intersects S either intersects S in a point or a simple closed curve [1], [16].

REFERENCES

[1] J. W. Alexander, " On the subdivisions of 3-spaces by a polyhedron," Proc. Nat. Acad. Sci., U.S.A., 10 (1924) pp. 6-8.

[2] ______, "An example of a simply connected surface bounding a region which is not simply connected," Proc. Nat. Acad. Sci. U.S.A., 10 (1924) pp. 8-10.

[3] W. R. Alford, "Some 'nice' wild 2-spheres in E^3," Topology of 3-Manifolds and Related Topics, Prentice Hall, 1962, pp. 29-33.

[4] L. Antoine, "Sur l'homéomorphie de deux figures et leurs voisinages," J. Math. Pures Appl., 4 (1921) pp. 221-325.

[5] B. J. Ball, "The sum of two solid horned spheres," Ann. of Math., (2) 69 (1959) pp. 253-257.

[6] R. H. Bing, "A homeomorphism between the 3-sphere and the sum of two solid horned spheres," Ann. of Math. (2) 56 (1952) pp. 354-362.

[7] ______, "Locally tame sets are tame," Ann. of Math. (2) 59 (1954) pp. 145-158.

[8] ______, "Approximating surfaces with polyhedral ones," Ann. of Math. (2) 65 (1957) pp. 456-483.

[9] ______, "An alternative proof that 3-manifolds can be triangulated," Ann. of Math. (2) 69 (1959) pp. 37-65.

[10] ______, "Conditions under which a surface in E^3 is tame, Fund. Math., 47 (1959) pp. 105-139.

[11] ______, "A surface is tame if its complement is 1-ULC," Trans. Amer. Math. Soc., 101 (1961) pp. 294-305.

[12] ______, "A wild surface each of whose arcs is tame," Duke Math. J., 28 (1961) pp- 1-16.

[13] ______, "Each disc in E^3 contains a tame arc," Amer. J. Math., 84 (1962) pp. 583-590.

[14] ______, "Approximating surfaces from the side," Ann. of Math. (2) 77 (1963) pp. 145-192.

[15] ______, "Pushing a 2-sphere into its oomplement," Mich. Math. J., 11 (1964) pp. 33-45.

[16] ______, "Spheres in E^3," Amer. Math. Monthly, 71 (1964) pp. 353-364.

[17] ______, "Improving the side approximating theorem," Trans. Amer. Math. Soc., 116 (1965) pp. 511-525.

[18] Morton Brown, "A proof of the generalized Schoenflies Theorem," Bull. Amer. Math. Soc., 66 (1960) pp. 74-76.

[19] ______, "Locally flat imbeddings of topological manifolds," Ann. of Math. (2) 75 (1962) pp. 331-341.

[20] C. E. Burgess, "Properties of certain types of wild surfaces in E^3," Amer. J. Math., 86 (1964) pp. 325-338.

[21] ______, "Characterizations of tame surfaces in E^3," Trans. Amer. Math. Soc., 114 (1965) pp. 80-97.

[22] ______, "Pairs of 3-cells with intersecting boundaries in E^3," Amer. J. Math., 88 (1966).

[23] L. O. Cannon, "Sums of solid horned spheres," Trans. Amer. Math. Soc., (to appear).

[24] B. G. Casler, "On the sum of two solid horned spheres," Trans. Amer. Math. Soc., 116 (1965) pp. 135-150.

[25] P. H. Doyle and J. G. Hocking, "Some results on tame discs and spheres in E^3," Proc. Amer. Math. Soc., 11 (1960) pp. 832-836.

[26] M. K. Fort, Jr., "A wild sphere which can be pierced at each point by a straight line segment," Proc. Amer. Math. Soc., 14 (1963) pp. 994-995.

[27] R. Fox and E. Artin, "Some wild cells and spheres in three-dimensional space," Ann. of Math. (2) 49 (1948) pp. 979-990.

[28] David S. Gillman, "Note concerning a wild sphere of Bing," Duke Math. J., 31 (1964) pp. 247-254.

[29] ______, "Side approximation, missing an arc," Amer. J. Math., 85 (1963) pp. 459-476.

[30] W. Graeub, Semilinear Abbildungen, Sitz.-Ber.d. Akad. d. Wissensch Heidelberg, 1950, pp. 205-272.

[31] H. C. Griffith, "A characterization of tame surfaces in three space," Ann. of Math. (2) 69 (1959) pp. 291-308.

[32] O. G. Harrold, Jr., "Locally tame curves and surfaces in three-dimensional manifolds," Bull. Amer. Math. Soc., 63 (1957) pp. 293-305.

[33] ______, "Locally peripherally unknotted surfaces in E^3," Ann. of Math. (2) 69 (1959) pp. 276-290.

[34] ______, and E. E. Moise, "Almost locally polyhedral spheres, " Ann. of Math. (2) 57 (1953) pp. 575-578.

[35] John Hempel, "A surface in S^3 is tame if it can be deformed into each complementary domain," Trans. Amer. Math. Soc., 111 (1964) pp. 273-287.

[36] Norman Hosay, "Conditions for tameness of a 2-sphere which is locally tame modulo a tame set," Amer. Math. Soc., Notices, 9 (1962) p. 117.

[37] ______, "Some sufficient conditions for a continuum on a 2-sphere to lie on a tame 2-sphere," Amer. Math. Soc., Notices, 11 (1964) pp. 370-371.

[38] L. D. Loveland, "Tame surfaces and tame subsets of spheres in E^3," Trans. Amer. Math. Soc. (to appear).

[39] ______, "Tame subsets of spheres in E^3," Pacific J. Math. (to appear).

[40] Joseph Martin, "A rigid sphere," Fund. Math. (to appear).

[41] E. E. Moise, "Affine structures in 3-manifolds, I-VIII," Ann. of Math. (2): 54(1951) pp. 506-533; 55 (1952) pp. 172-176; 55(1952) pp. 203-214; 55 (1952) pp. 215-222; 56 (1952) pp. 96-114; 58 (1953) pp. 107 ; 58 (1953) pp. 403-408; 59 (1954) pp. 159-170;

[42] C. D. Papakyriakopoulos, "On Dehn's Lemma and the asphericity of knots," Ann. of Math., 66 (1957)pp. 1-26.

[43] A. Shapiro and J. H. C. Whitehead, "A proof and extension of Dehn's Lemma," Bull. Amer. Math. Soc., 64 (1958) pp. 174-178.

[44] John R. Stallings, "Uncountably many wild disks," Annals of Math. (2) 71 (1960) pp. 185-186.

ADDITIONAL QUESTIONS ON 3-MANIFOLDS

1. (R. H. Bing) Can Dehn's Lemma be generalized in either of the following two directions?

   a) Does a simple closed curve bound a disc in a 3-manifold if it can be shrunk to a point in its own complement?

   b) Suppose f is a piecewise linear map of E^2 into E^3 which takes unbounded sets onto unbounded sets, D is a disc in E^2 such that f^{-1} is a homeomorphism on $f(D)$, and N is a neighborhood of the singularities of f. Is there a homeomorphism h of E^2 onto a closed set in $f(E^2) \cup N$ which agrees with f on D ?

2. (David Henderson) Let D be a disc-with-holes and f a map of $(D, \text{Bd } D)$ into $(M^3, \text{Bd } M^3)$, where M^3 is a 3-manifold. We say that f is <u>non-pullable</u> if there does not exist a map $F\colon D \to \text{Bd } M^3$ such that $F|\text{Bd } D = f|\text{Bd } D$.

   Conjecture: If M is an orientable 3-manifold, D a disc-with-holes, and $f\colon (D, \text{Bd } D) \to (M, \text{Bd } M)$ is non-pullable, then there exists a disc-with-holes D' and an embedding f' such that $f'\colon (D', \text{Bd } D') \to (M, \text{Bd } D)$ is non-pullable.

Note that if D is a disc, then this conjecture reduces to the loop theorem and Dehn's Lemma. Using results of Shapiro and Whitehead [1], it is enough to get a map f' such that is non-pullable and is an embedding when restricted to a neighborhood of $\text{Bd } D'$. An affirmative answer to this conjecture would give an affirmative answer to Question 13 of [2] and would be a useful tool in "scissors and paste" topology.

References:

1. Shapiro and Whitehead, "A proof and extension of Dehn's Lemma," Bull. Amer. Math. Soc., vol. 64 (1958) pp. 174-178.
2. Sumner Institute on Set-theoretic Topology, Madison, Wisconsin, 1955.

3. (R. H. Bing) Which of the following conditions is enough to imply that a 2-sphere S in E^3 is tame?

   (a) S can be approximated from either side by a singular 2-sphere [1].

   (b) S can be touched from either side by a pencil [2].

   (c) At each point S lies locally between two tangent 2-spheres [2].

(d) Each horizontal cross-section of S is connected.

(e) Each horizontal cross-section of S is either a point or a simple closed curve [2].

(f) For each $\varepsilon > 0$ there is a $\delta > 0$ such that each simple closed curve of S of diameter less than δ can be shrunk to a point on ε-sets on either side of S.

(g) S can be pierced along each arc in it by a tame disc [1].

(h) S is homogeneous under a space homeomorphism [1].

(i) S is tame mod a closed set X such that X lies in a plane and has no degenerate components. (Hosay's thesis gives some partial solutions.)

References:

1. Bing, R. H., "Embedding surfaces in 3-manifolds," Proc. International Congress of Mathematicians, 1962, pp. 457-458.

2. ______, "Spheres in E^3," Amer. Math. Monthly, vol. 71 (1964) pp. 353-364.

4. (R. McMillan) If M is a compact 3-manifold with boundary (non-empty) and h is a homeomorphism of M onto itself such that $h|\text{Bd } M$ is the identity, what conditions on h are sufficient to insure that h is isotopic to the identity?

5. (F. Burton Jones) Suppose that M is a non-separating subcontinuum of S^2. Then if a and b are points of M which are not cut from each other in M by any zero dimensional closed subset of M, then there is a closed disc in M which contains a and b. [Jones, about 1960.]

What are possible true extensions to S^3? For example: If M is a point-like subcontinuum of S^3 and no 1-dimensional closed subset of M cuts a from b in M, does M contain a locally connected subcontinuum K containing both a and b such that no 1-dimensional closed subset of K separates a from b in K ?

6. (R. H. Bing) Suppose $h: E^3 \to E^3$ is a periodic map whose fixed point set is a straight line. Is h equivalent to a rotation about E^1?

7. (Robert Craggs) Can each periodic transformation of S^3 onto S^3 be approximated by a piecewise linear periodic transformation? If the answer is yes, must the fixed point set for the approximating periodic transformation be the same type of sphere as the fixed point set of the original periodic transformation?

8. (R. H. Bing) If $h: E^3 \to E^3$ is a periodic map with a point as the fixed point set, is h the composition of two periodic maps h_1 and h_2 where h_1 is a rotation about a line (possibly wild) and h_2 is a reflection through a plane (possibly wild)?

CHAPTER II: THE POINCARÉ CONJECTURE

HOW NOT TO PROVE THE POINCARÉ CONJECTURE

by

John Stallings*

Introduction

I have committed—the sin of falsely proving Poincaré's Conjecture. But that was another country; and besides, until now no one has known about it.

Now, in hope of deterring others from making similar mistakes, I shall describe my mistaken proof. Who knows but that somehow a small change, a new interpretation, and this line of proof may be rectified!

In the back of my mind when I conceived my proof was this theorem.

THEOREM 0: (For $n \neq 2$). Let $f\colon M \to K$ be a map of a connected orientable n-manifold into an n-complex, and let $C_1, \ldots, C_k$ be some of the n-simplexes of K such that the degree of f on each C_i is zero (that is, the homology map induced by f, $H_n(M) \to H_n(K, K\text{-int } C_i)$, is zero). Suppose f induces a homomorphism of $\pi_1(M)$ onto $\pi_1(K)$. Then f is homotopic to a map into $K - (\text{int } C_1 \cup \ldots \cup \text{int } C_k)$.

A special argument establishes this theorem for $n = 1$. For $n \geq 3$ we may argue as follows. We shall make a number of changes on f which will be independent of each other, since $n \geq 3$; hence we need only consider the case that $k = 1$ and suppose that C_1 is covered twice with opposite orientations by $f(M)$. The inverse image of a small cell in C_1 is the union of two cells A and B in M. Let P be a path in M from A to B; fP represents an element of $\pi_1(K)$. Since $\pi_1(M) \to \pi_1(K)$ is onto, we can modify P by adding on a loop whose image represents the inverse of fP; thus we can suppose fP is a null-homotopic loop in K.

Since the dimension of M is at least 3, we can choose P to be a non-singular path and change f by a homotopy in the neighborhood of P so that $f(P) \subset C_1$. If T is a tube around P, then $A \cup T \cup B$ will be an n-cell mapped into C_1 with degree zero. A further homotopy within $A \cup T \cup B$

---

* Sloan Foundation Fellow

will uncover a point of C_1; by pushing away from that point, we uncover al of int C_1.

But, in my proof of Poincaré's Conjecture, I need this theorem for $n = 2$. The argument above fails in this case for several reasons. We canno uncover 2-cells independently of each other; we cannot make the path P non-singular; if P were non-singular, the homotopy bringing $f(P)$ into C_1 might cause us to cover up cells which we want to uncover.

The reader may be able to patch up some of these points. If he patches up all these points, he will have proved the Poincaré Conjecture (for we shall show how Theorem 0 for $n = 2$ implies the Poincaré Conjecture incorrectly. For, Theorem 0 is false for $n = 2$: Consider a torus with two 2-cells C_1 and C_2 attached to kill the fundamental group; there is a map of the 2-sphere into this complex; by a homotopy we can uncover either C_1 o C_2, but not both simultaneously.

1. A conjecture about the 3-sphere

In the 3-sphere S^3 let T be a tame 2-manifold such that both co ponents of $S^3 - T$ have free fundamental groups. Let U and V denote the closures of the components of $S^3 - T$.

According to theorems of Papakyriakopoulos, both U and V are handlebodies. The only conceivable such embedding of T is shown in Fig. 1

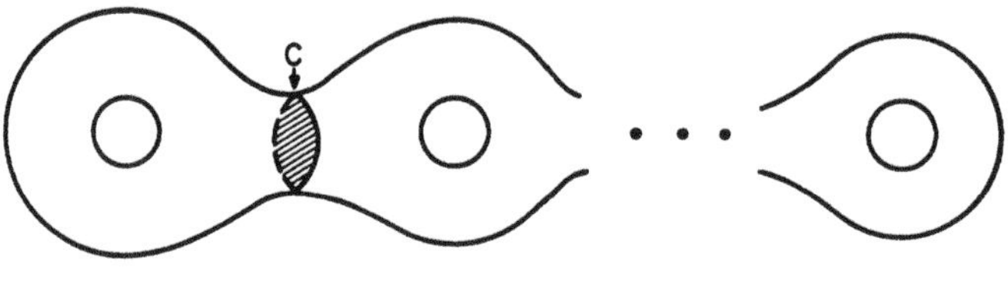

Fig. 1

However, if the genus of T is greater than 1, this standard embedding has a property which does not seem to follow immediately from the fact that U and V are handlebodies. Namely, there is a simple closed curve C on T, not contractible on T, yet bounding 2-cells both in U and in V.

CONJECTURE A: The existence of such a curve C can be proved only from the hypothesis that both U and V are handlebodies.

If we hope that Conjecture A is true, a reasonable direction to attempt a proof of the Poincaré Conjecture can be made as follows.

Poincaré's Conjecture is that any simply-connected 3-manifold M is a 3-sphere. It is known that any orientable 3-manifold such as M, has a Heegaard representation as $U \cup V$, where U and V are handlebodies and $U \cap V$ is their common boundary, a 2-manifold T.

If the genus of T should happen to be one, then M is a lens space, and so, if simply-connected, is a 3-sphere.

Assume that we could prove Conjecture A for M, rather than for the 3-sphere: That is, if the genus of T is greater than one, then on T there is a simple closed curve C, not contractible on T, yet bounding 2-cells in both U and V. Then we could write M as the connected sum $M_1 \# M_2$ of two manifolds whose Heegaard representations would have less genus. And so by induction on the genus, we would know that M is indeed a 3-sphere.

2. Reduction to group theory

Let $M = U \cup V$, $T = U \cap V$ be a Heegaard representation of a 3-manifold. We obtain a diagram of fundamental groups, with homomorphisms induced from inclusions:

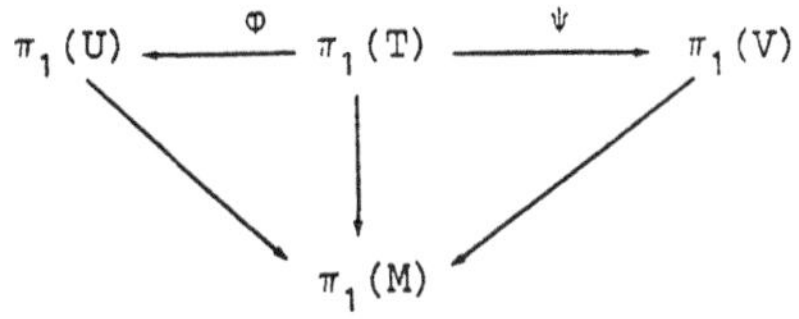

Since φ and ψ are homomorphisms onto, it follows from van Kampen's Theorem that $\pi_1(M)$ is isomorphic to the quotient of $\pi_1(T)$ by $(\ker \varphi) \cdot (\ker \psi)$. Hence M is simply connected exactly when

$$\pi_1(T) = (\ker \varphi) \cdot (\ker \psi) \quad .$$

We have an obvious homomorphism to investigate,

$$\varphi \times \psi \colon \ \pi_1(T) \to \pi_1(U) \times \pi_1(V) \quad .$$

The kernel of this homomorphism is clearly,

$$\ker \varphi \times \psi = (\ker \varphi) \cap (\ker \psi) \quad .$$

Therefore our geometric problem has been "reduced," by virtue of Dehn's Lemma, to the more algebraic problem: Does $\ker \varphi \times \psi$ contain an element which can be represented by a simple closed curve on T ?

Now this cannot be true for arbitrary 3-manifolds, for if it were we would have proved, "Every 3-manifold is a connected sum of lens spaces," which is absurd.

THEOREM 1: $\varphi \times \psi$ is a homomorphism onto, if and only if M is simply connected.

First, if $\varphi \times \psi$ is onto, since $\pi_1(U) \times \pi_1(V)$ is the product of the kernels of the projections onto its factors, it follows that $\pi_1(T)$ is the product of $\ker \varphi$ and $\ker \psi$, and hence M is simply connected.

Conversely, if M is simply connected, then $\pi_1(T) = (\ker \varphi)\cdot(\ker \psi)$. Let (α, β) be an arbitrary element of $\pi_1(U) \times \pi_1(V)$. Since φ and ψ are onto, there are α_1, β_1 in $\pi_1(T)$ such that $\varphi(\alpha_1) = \alpha$ and $\varphi(\beta_1) = \beta$. We can decompose α_1 and β_1 thus: $\alpha_1 = x\alpha_2$, $\beta_1 = \beta_2 y$, where x and β_1 belong to $\ker \varphi$, and α_2 and y belong to $\ker \psi$. Then $\varphi \times \psi(\alpha_2 \cdot \beta_2) = (\alpha, \beta)$. Hence $\varphi \times \psi$ is onto.

Now our "reduction" can be stated as the following conjecture:

CONJECTURE B: Let T be an orientable 2-manifold of genus $n > 1$. Let F_1 and F_2 be free groups of rank n. Let $\eta\colon \pi_1(T) \to F_1 \times F_2$ be a homomorphism onto. Then there is a non-trivial element of $\ker \eta$ which is represented by a simple closed curve on T.

We have shown that Conjecture B implies both the Poincaré Conjecture and Conjecture A. It is likely that from the data of Conjecture B we can reconstruct a 3-manifold; in which case, then, conversely, the Poincaré Conjecture and Conjecture A together would imply Conjecture B.

3. Finding simple closed curves.

Papakyriakopoulos and Maskit have discovered an interesting characterization of simple closed curves in terms of planar covering spaces. Their results show that Conjecture B is equivalent to the following:

CONJECTURE C: In the situation of Conjecture B, there is a non-trivial normal subgroup N of $\pi_1(T)$, such that $N \subset \ker \eta$ and such that the covering space of T which corresponds to N is a planar surface.

However, in our discussion we are interested in a different characterization of simple closed curves.

Let us say that a homomorphism $\varphi\colon G \to A * B$ of a group G into a free product of groups is <u>essential</u> if there is no element $x \in A * B$ such that $x \cdot \varphi(G) \cdot x^{-1}$ is contained in one of the factors A or B.

THEOREM 2: If $G = \pi_1(T)$, where T is a closed 2-manifold, and if $\varphi\colon G \to A * B$ is an essential homomorphism, then there is some non-trivial element of $\ker \varphi$ which is represented by a simple closed curve on T.

<u>Proof</u>: Represent $A * B$ as the fundamental group of $X = X_A \cup X_B$ where X_A and X_B are open sets in X with fundamental groups A and B

respectively, and where $X_A \cap X_B$ is simply connected. Represent φ by a continuous function $f: T \to X$.

We can divide T into submanifolds T_A and T_B, whose intersection is their common boundary, such that $f(T_A) \subset X_A$ and $f(T_B) \subset X_B$. The components of $T_A \cap T_B$ are simple closed curves, whose images by f lie in $X_A \cap X_B$ and hence are contractible in X.

If, contrary to the conclusion we wish to draw, every such simple closed curve were trivial on T, then any one, say C, would bound a 2-cell D. Redefine f on D so as to map D into $X_A \cap X_B$, and then redefine T_A and T_B; this will reduce the number of components of $T_A \cap T_B$. Finally, we obtain a map f' and a division of T so that either T_A or T_B is empty. f' induces the same homomorphism on fundamental groups as f does, modulo an inner automorphism (we may have moved the basepoint around), so that φ is inessential, contradicting hypothesis.

Reinterpreting Conjecture B in the light of this theorem, we have:

CONJECTURE D: In the situation of Conjecture B, the map $\eta: \pi_1(T) \to F_1 \times F_2$ can be factored through an essential map of $\pi_1(T)$ into some free product $A * B$.

Thus have we replaced the purely geometric Poincaré Conjecture by the purely algebraic Conjecture D.

4. Geometric "proof" of Conjecture D

Let us consider the map $\eta: \pi_1(T) \to F_1 \times F_2$ with components φ and ψ, so that $\eta(x) = (\varphi(x), \psi(x))$. It is possible given that η is onto, after a moderate amount of algebraic slickness, to find a presentation of $\pi_1(T)$ as $\{a_1, b_1, \ldots, a_n, b_n: \Pi_{i=1}^{n} [a_i, b_i] = 1\}$ and bases $\{\alpha_1, \ldots, \alpha_n\}$ and $\{\beta_1, \ldots, \beta_n\}$ of F_1 and F_2, such that, <u>modulo the commutator subgroups</u>, $\eta(\alpha_i) \equiv (\alpha_i, 1)$ and $\eta(\beta_i) \equiv (1, \beta_i)$.

Now interpret F_1 as the fundamental group of a bouquet of circles $X_1 \vee X_2 \vee \ldots \vee X_n$, where X_i corresponds to the basis element α_i. Similarly interpret F_2 as the fundamental group of $Y_1 \vee \ldots \vee Y_n$, with Y_i corresponding to β_i. Define $W = (X_1 \cdots X_n) \times (Y_1 \cdots Y_n)$.

We may interpret η as the homomorphism induced from a function $f: T \to W$. Because of our clever choice of bases, f will have degree zero on the tori $X_i \times Y_j$ for $i \neq j$ and degree one on the tori $X_i \times Y_i$.

Let W^* denote $(X_1 \times Y_1) \cup (X_2 \times Y_2) \cup \ldots \cup (X_n \times Y_n)$. This is the union of n tori with a single point in common.

If only, alas, Theorem 0 were valid for dimension two, we could conclude that the map $f: T \to W$ could, up to homotopy, be factored through a map $g: T \to W^*$. The map g would have to induce an isomorphism on 1-dimensional homology, and hence, writing $\pi_1(W^*)$ as the obvious free product, the map of fundamental groups would be essential.

And so Conjecture D would be proved. Poincaré's Conjecture would follow. Fame and Fortune would be ours.

5. Conclusion.

There are two points about this incorrect proof worthy of note.

The first is that when we try to prove Theorem 0 in dimension two, we always run up against the problem of trying to simplify, by some geometric trick, the situation. But any little homotopy that would simplify the picture always in fact, greatly complicates it. This phenomenon has characterized every attempt that I have made or heard of to prove Poincaré's Conjecture. This is the place to look for flaws in any asserted "proof."

The second point is that I was unable to find flaws in my "proof" for quite a while, even though the error is very obvious. It was a psychological problem, a blindness, an excitement, an inhibition of reasoning by an underlying fear of being wrong. Techniques leading to the abandonment of such inhibitions should be cultivated by every honest mathematician.

REFERENCES

[1] B. Maskit, "A theorem on planar covering surfaces with applications to 3-manifolds," Ann. of Math., 81 (1965) pp. 341-355.

[2] C. D. Papakyriakopoulos, "A reduction of the Poincaré Conjecture to group-theoretic conjectures," Ann. of Math., 77 (1963) pp. 250-305.

MAPPING A 3-SPHERE ONTO A HOMOTOPY 3-SPHERE

by

R. H. Bing*

1. Introduction

We shall use S^3 to denote a 3-sphere—that is a set homeomorphic to the one point compactification of Euclidean 3-space. It may be convenient to regard S^3 as the union of two cubes sewed together with a homeomorphism on their boundaries. We suppose that S^3 is triangulated.

We shall use M^3 to denote a homotopy 3-sphere—that is a compact connected simply connected 3-manifold. We also suppose it is triangulated.

The Poincaré conjecture is that M^3 is topologically equivalent to S^3. See [1] for a discussion of attempts before 1964 to solve the conjecture. There were serious attempts made during 1964 to prove the Poincaré conjecture in the affirmative. For a while during the summer of 1964 it was felt by some that Wolfgang Haken or (and) Valentin Poénaru would be able to construct an airtight proof. Their plans of attack were essentially the same although Haken used a piecewise linear approach and Poénaru a differentiable approach.

This common approach consisted of removing the interior of a 3-simplex from M^3 so as to have left a compact connected, simply connected 3-manifold with boundary whose boundary is a 2-sphere. We denote this remaining homotopy 3-cell by K^3 and call it a fake cube. It follows from a theorem of Hurewicz that since $\pi_1(K^3) = 0$, $H_2(K^3) = 0$, then $\pi_2(K^3) = 0$. Hence, there is a homotopy pulling the boundary of K^3 to a point in K^3. Both Haken and Poénaru try to make this homotopy "nice" so as to conclude that K^3 is a real cube. Neither succeeded.

Suppose $H_t\colon \mathrm{Bd}\, K^3 \to K^3$ $(0 \leq t \leq 1)$ is a homotopy of $\mathrm{Bd}\, K^3$ into K^3 such that H_0 is the identity map and H_1 is a constant map. It can be shown that if each H_t $(0 \leq t \leq 1)$ takes $\mathrm{Bd}\, K^3$ homeomorphically into a tame 2-sphere in $\mathrm{Int}\, K^3$, then K^3 is a topological cube.

As far as I know, the proof has not appeared in print but the approach is to show that if $0 < t_0 \leq 1$, then $\mathrm{Bd}\, K^3$ and $H_{t_0}(\mathrm{Bd}\, K^3)$ co-

* Work on this paper was supported by the National Science Foundation under Grant GP-3857.

bound a hollow ball (3-dimensional annulus) in K^3.

Question 1. Can one prove that K^3 is a real cube if one merely assumes that each H_t $(0 \leq t \leq 1)$ takes Bd K^3 homeomorphically into a 2-sphere (perhaps wild) in Int K^3?

(After this talk was given, Robert F. Craggs answered the above question in the affirmative by using the result of McMillan [4] that if a 2-sphere is embedded in a 3-manifold M^3 then there is a hollow ball with handles in M^3 such that the two sphere separates in M^3 the two boundary components of the hollow ball with handles.)

Instead of pushing the boundary of a fake cube, one might try pushing spanning arcs to near the boundary. It follows from techniques used in [1] that a fake cube K^3 is a topological 3-cell if for each polyhedral spanning arc A^1 of K^3 there is an isotopy H_t: $A^1 \to K^3$ $(0 \leq t \leq 1)$ such that H_0 is the identity, H_t is the identity on the ends of A^1, $H_1(A^1)$ lies in a Cartesian product neighborhood of Bd K^3, and each $H_t(A^1)$ is a tame arc.

Question 2. Can one prove that K^3 is a real cube if one drops the assumption that each $H_t(A^1)$ is tame?

Although neither Haken nor Poénaru got a solution to the Poincaré conjecture, each was able to salvage some good results out of his effort. An interesting result obtained by Haken was the following:

THEOREM 1. There is a map f of S^3 onto M^3 such that the image of the singularities of f lie on the interior of a cube with handles X^3.

A point $p \in S^3$ is a nonsingular point of f if there is a neighborhood U of $f(p)$ in $f(S^3)$ such that f^{-1} is a homeomorphism on U. Other points of S^3 are singular points of f. A cube with handles is a set homeomorphic to the union of a finite number of 3-balls C^3, $H_1{}^3$, $H_2{}^3$, ..., $H_n{}^3$ such that $C^3 \cap H_i{}^3 = \text{Bd } C^3 \cap \text{Bd } H_i{}^3$ is the union of two mutually exclusive disks and no two of the handles $H_i{}^3$ intersect each other.

A stronger form of Theorem 1 was stated by Moise in [3]. However, we do not have Moise's proof so we will model ours after Haken's given in [2].

2. Three Lemmas

We shall list three lemmas that can be used to help prove Theorem 1.

Consider the Cartesian product of a disk with an arc. If the disk is round and the arc is short straight segment, the resulting topological cube resembles a pill box. Figure 1 shows two pill boxes P^3, Q^3 sewed together along a disk S^2 on the lateral side of each.

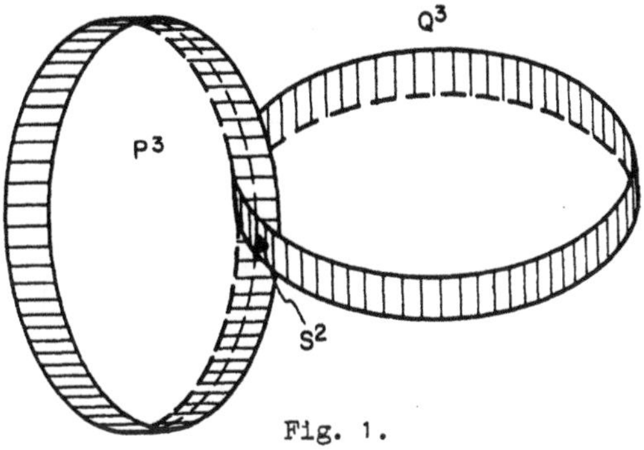

Fig. 1.

Pill Box Lemma 1. There is a homeomorphism of $P^3 \cup Q^3$ of Fig. 1 onto a ball that takes the two bases of P^3 and the exposed part of the lateral side of Q^3 onto the upper hemisphere of the ball and the two bases of Q^3 and the exposed part of the lateral side of P^3 onto the lower hemisphere.

We will not include a proof of the Pill Box Lemma. We note that the simple closed curve on $P^3 \cup Q^3$ corresponding to the equator of the ball is the union of four circles each with a gap cut in it where P^3 intersects Q^3.

We mention where the Pill Box Lemma is used. Suppose one is using the approach of Haken or Poénaru and pushing a two sphere in toward a 2-complex in M^3. A regular neighborhood of this 2-complex resembles the union of a tubular neighborhood T^3 of the 1-skeleton of the 2-complex together with some pill boxes P_i^3 sewed onto T^3 to cover the middle portion of the 2-simplexes of the 2-complex as shown in Fig. 2. If the 2-sphere has been pushed in to cover the bases of the P_i^3's and the part Bd T^3 not on the lateral edge of a P_i^3 then one can push in further across the double pill box shown in Fig. 2 so that the part of the sphere that once went onto the bases of P_i^3 and the lateral edge of the right pill box now go onto the lateral edges of P_i^3 and the bases of the right pill box. Such a push is cometimes called a baseball move since the part of the double pill box corresponding to the equator of a ball is in somewhat the same position as the seams on a baseball. We have essentially replaced one of the two halves of a baseball cover with the other.

If the P_i^3 of Fig. 2 is added to T^3, then the new T^3 and the new S^3 - Int T^3 each has one fewer handle.

The second lemma we shall find useful is the following:

LEMMA 2. Suppose G^1 is a finite polygonal graph that can be pulled to a point in a triangulated 3-manifold M^3 in which it lies. Then there is a polyhedral cube with handles T^3 in M^3 containing G^1 on its interior such that G^1 can be shrunk to a point in T^3.

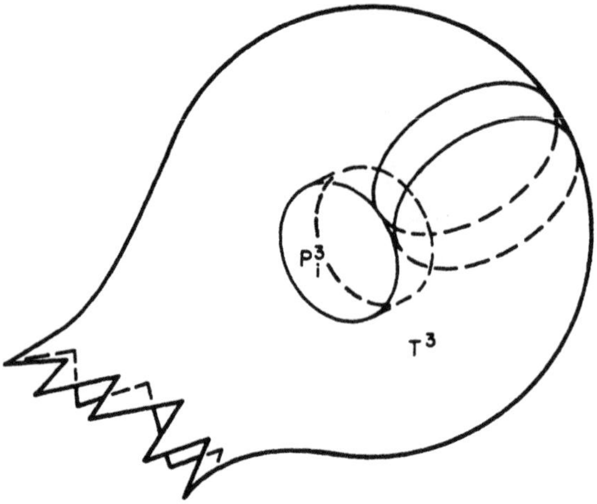

Fig. 2

With no loss of generality we suppose that G^1 is connected. In this connected G^1 a maximal tree can be shrunk to a point such that G^1 becomes a bouquet of simple closed curves. In Section 6 of [2] Haken shows this bouquet of simple closed curves bounds a singular bouquet of disks whose regular neighborhood is a cube with handles.

LEMMA 3. Any orientable compact, connected, 3-manifold N^3 with non-null boundary is the union of a cube with handles T^3 and a finite number of 3-cells $P_1^3, P_2^3, \ldots, P_n^3$ such that each $P_i^3 \cap T^3$ is an annulus on $\text{Bd } P_i^3 \cap \text{Bd } T^3$.

This is a lemma used by McMillan in [4].

We suppose N^3 is triangulated and T^1 is the 1-skeleton of the triangulation. A tubular neighborhood of T^1 is a cube with handles. One can obtain this tubular neighborhood of T^1 as a remainder if one bores holes in M^3 along the dual 1-skeleton of the triangulation. These holes correspond to the P_i^3's. Although as originally drilled, a hole may have its ends on the lateral sides of previous holes rather than on $\text{Bd } N^3$, an isotopy remedies this situation.

Another way of proving Lemma 3 involves collapsing N^3 to a 2-spine. A tubular neighborhood of the 1-skeleton of the spine provides us with T^3 and a thickening of the middle part of the 2-simplexes of the spine provides us with the P_i^3's. In this case the P_i^3's resemble pill boxes. In case N^3 is a fake cube, the 2-spine is a contractible 2-complex such as shown in Fig. 3. It shows a house-with-two rooms, an igloo shaped figure, and a dunce's cap strung together.

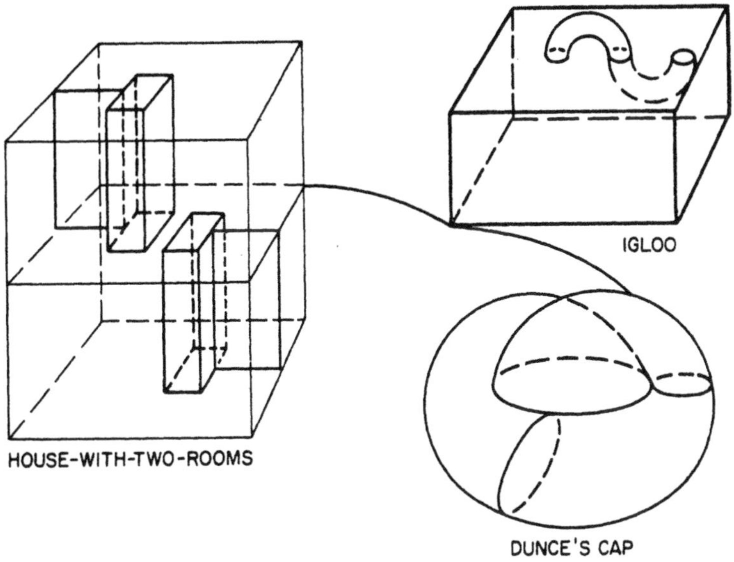

Fig. 3

A fake cube is the contractible union of a cube with handles and a finite number of mutually exclusive pill boxes each attached to the cube with handles along an annulus. The fundamental group of the cube with handles is a free group on n generators where n is the number of handles. Each pill box provides a relation so the fundamental group of the fake cube is

$$G = \{a_1, a_2, \ldots, a_n / r_1 = r_2 = \cdots = r_n = 1\} .$$

Since a fake cube is simply connected, the fundamental group G for a fake cube is trivial.

<u>Question</u> 3. What are some exotic presentations of the trivial group of the following sort

$$G = \{a_1, a_2, \ldots, a_n / r_1 = r_2 = \cdots = r_n = 1\}?$$

An example of such a presentation is

$$G = \{a, b / a^i b^j = a^r b^s = 1 \quad \text{where} \quad |is - rj| = 1\} .$$

This one does not lead to a counterexample to the Poincaré conjecture because it is possible to obtain a real cube by adding two pill boxes to a cube with handles so that the pill boxes yield the suggested relations. Perhaps some other group does lead to a counterexample.

3. Proof of Theorem 1

The map f is defined in three parts. First we define it outside a 3-ball B^3. Then it is defined on some pits in B^3. Finally, it is defined on the part of B^3 not in these pits. See Fig. 4.

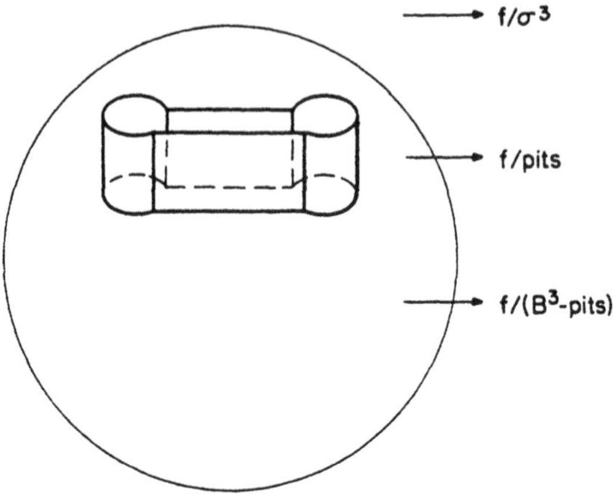

Fig. 4

<u>Part</u> 1. Let f be the map of a 3-simplex σ^3 of S^3 onto a 3-simplex of M^3. We let B^3 be the 3-cell S^3 - (Int σ^3) and K^3 be the fake cube M^3 - f(Int σ^3). Our task now is to extend f to B^3.

<u>Part</u> 2. It follows from Lemma 3 that K^3 is the union of a cube with handles T^3 and a finite number of mutually exclusive pill boxes $P_1^3, P_2^3, \ldots, P_n^3$ sewed onto T^3 along their lateral sides. It follows from Lemma 2 that T^3 lies in a cube with handles X^3 in M^3 such that T^3 can be shrunk to a point in Int X^3 and some open subset U^3 of f(σ^3) misses X^3.

Let A_i^2 be the lateral side of P_i^3 and S_i^2 be a disk on A_i^2 intersecting each end of P_i^3 in an arc. Let R_i^2 be a disk on Bd $T^3 - \cup$ Int A_j such that $R_i^2 \cap (P_j^3 \cup R_j^2) = 0$ $(i \neq j)$ and $R_i^2 \cap P_i^3 = S_i^2$ - Int A_i. See Fig. 5.

If $R_i^2 + S_i^2$ were the lateral edge of a pill box Q_i^3 in T^3, $P_i^3 \cup Q_i^3$ would be a double pill gox as mentioned in Lemma 1. There is no assurance that this is the case. However, since $R_1^2 \cup S_1^2$ lies in T^3, which in turn can be shrunk to a point in Int X^3 there is a map g_1 of

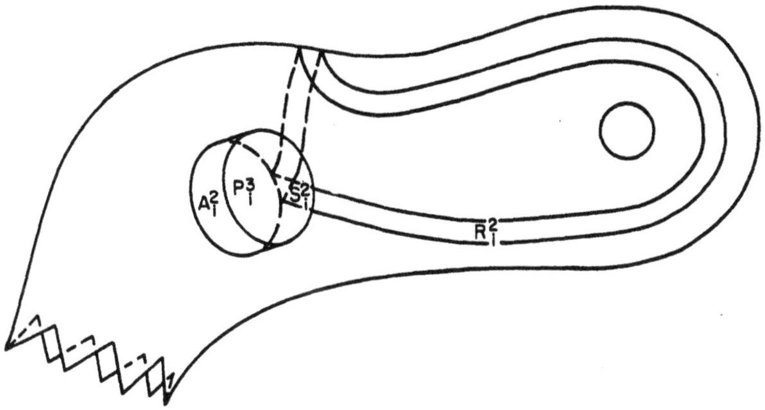

Fig. 5

the double pill box $P^3 \cup Q^3$ of Lemma 1 such that g_1 takes P^3 homeomorphically onto P_1^3, the later sides of P^3 onto A_1^2, the lateral side of Q^3 homeomorphically onto $S_1^2 \cup R_1^2$ and Q^3 into Int X^3 (perhaps singularly).

Let $B_1^2 = f^{-1}(R_1^2 \cup \text{ends of } P_1^3)$. It is a disk on Bd B^3 resembling two disks joined by a narrow band. Let $C_1^3, C_2^3, \ldots, C_n^3$ be mutually exclusive 3-cells in B^3 such that $C_1^3 \cap \text{Bd } B^3 = B_1^2$. Let h_1 be a homeomorphism of C_1^3 onto $P^3 \cup Q^3$ of Lemma 1 such that $g_1 h_1 / B_1^2 = f / B_1^2$. Note that $f(\text{Bd } C_1^3 - B_1^2) \subset X^3$. We define f on the pits C_1^3 by $f/C_1^3 = g_1 h_1$.

<u>Part</u> 3. Let $B^{3'}$ denote the 3-ball which is the closure of $B^3 - \cup C_1^3$. Note that $f(\text{Bd } B^{3'}) \subset X^3$. Since X^3 is a cube with handles, $\pi_2(X^3) = 0$. Hence $f/\text{Bd } B^{3'}$ can be extended to map of $B^{3'}$ into Int X^3.

It may be shown that $f\colon\ S^3 \to K^3$ is of degree 1 since f^{-1} is a homeomorphism on U^3. The only points of M^3 that have more than one inverse lie on a compact subset of Int X^3. Hence the image of the singularities of f lie in Int X^3.

4. Replacing cubes with holes by cubes with handles

Suppose C^3 is a polyhedral 3-cell and $C_1^3, C_2^3, \ldots, C_n^3$ is a collection of mutually exclusive polyhedral 3-cells in C^3 such that Bd $C^3 \cap$ Bd C_1^3 is the union of two mutually exclusive disks $A_1^2, A_1^{2'}$. We regard C_1^3 as $A_1^2 \times [0,1]$ where $A_1^2 \times 0 = A_1^2$ and $A_1^2 \times 1 = A_1^{2'}$.

Then $C^3 - \cup (\text{Int } A_i \times [0,1])$ is a <u>cube with holes</u> and each $\text{Int } A_i \times [0,1]$ is a <u>hole</u>.

Let p_i be a point of $A_i^{\,2}$. The hole $\text{Int } A_i^{\,2} \times [0,1]$ is <u>knotted</u> or <u>unknotted</u> according as $p_i \times [0,1]$ is a knotted or unknotted spanning arc of C^3. A tame spanning arc of C^3 is <u>unknotted</u> or <u>knotted</u> according as it, together with an arc on $\text{Bd } C^3$, does or does not bound a disk in C^3.

It is known that each polyhedron in S^3 bounded by a connected 2-manifold is topologically a cube with holes. Hence $f^{-1}(X^3)$ of Theorem 1 is a cube with holes and M^3 was obtained from S^3 by removing a cube with holes and replacing it with a cube with handles. Theorem 1 advises us that any homotopy 3-sphere can be obtained as such a simplification of S^3. Theorem 1 tells us a way of getting a counterexample to the Poincaré conjecture if there is one.

The following result is rather easy to verify.

THEOREM 2. Suppose $Y^{3'}$ is a tame cube with holes in S^3, g is a map $Y^{3'}$ onto a cube with handles Y^3 such that $g/\text{Bd } Y^{3'}$ takes $\text{Bd } Y^{3'}$ homeomorphically onto $\text{Bd } Y^3$. If one replaces $Y^{3'}$ by Y^3 in S^3 (the sewing being done by a g along $\text{Bd } Y^{3'}$), one obtains a homotopy 3-sphere.

Theorem 1 tells us that any homotopy 3-sphere can be obtained from S^3 by removing some cube with holes and replacing it with a cube with handles while Theorem 2 tells us that anytime this replacement is done in a certain way, one obtains a homotopy 3-sphere.

Hempel showed that if X is a cube with one hole, then there is a map f of X onto a cube with a handle such that f/Bd X takes Bd X homeomorphically onto Bd f(X). Whether or not each cube with holes in S^3 can be replaced in the "certain way" depends on the answers to the following questions.

<u>Question</u> 4. If X is a cube with two holes, is there a map f of X onto a cube with two handles such that f/Bd X takes Bd X homeomorphically onto the boundary of the cube with handles?

<u>Question</u> 5. If X is a cube with holes, under what conditions is there a map f of X onto a cube with handles such that f takes Bd X homeomorphically onto the boundary of the cube with handles?

A cube with a knotted hole is topologically different from a cube with an unknotted hole since one has a non-Abelian fundamental group and the other has a fundamental group which is infinite cyclic.

However, a cube with two holes one of which is knotted and the other of which is unknotted may be topologically equivalent to a cube with two straight holes. Figure 6 shows how to start with a cube with two straight holes, gradually move the bottom base of the right hole along parallel to the dotted line until the right hole becomes knotted.

Question 6. Suppose $C^3 - (\text{Int } A_1 \times [0,1]) - (\text{Int } A_2 \times [0,1])$ is a cube with two holes tamely embedded in E^3 and topologically equivalent to a cube with handles such that $\text{Int } A_1 \times [0,1]$ is unknotted but $\text{Int } A_2 \times [0,1]$ is knotted. In Fig. 6 the knotted hole is a trefoil knot. Other toroidal knots could have been similarly obtained. What kind of knots could have been obtained?

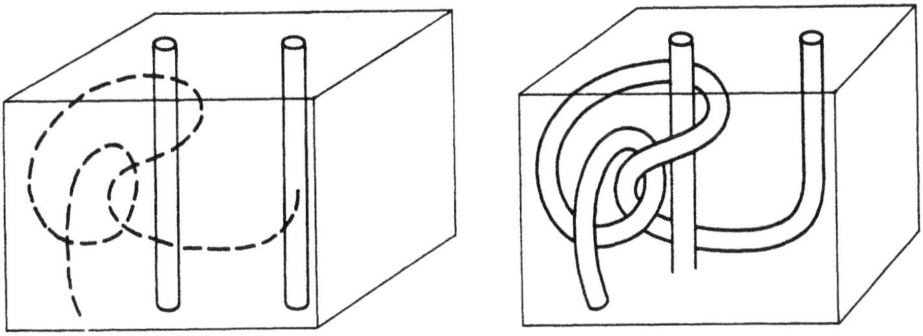

Fig. 6

One cannot tell at a glance whether or not a cube with holes is topologically equivalent to a cube with handles. Even if both holes are knotted, one cannot be sure it is topologically inequivalent to a cube with handles. For example if one winds the straight hole of Fig. 6 about the knotted hole, one obtains two holes as shown in the bottom part of Fig. 7.

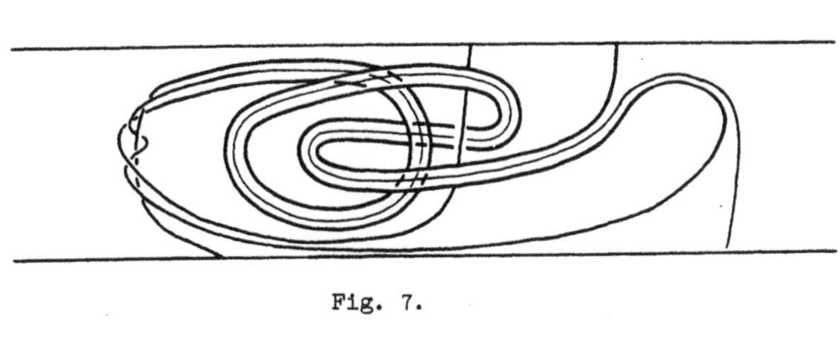

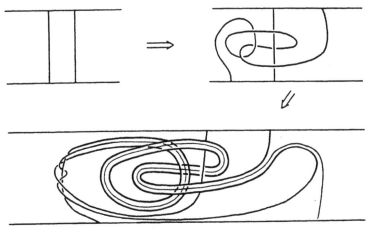

Fig. 7.

5. Simplifying handles

One might try to prove the Poincaré conjecture by simplifying either f or X^3 of Theorem 1.

THEOREM 3. In order for M^3 of Theorem 1 to be a counterexample to the Poincaré conjecture it is necessary that X^3 have more than one handle.

Proof: If X^3 has no handles, each of X^3 and M^3 - Int X^3 is a 3-cell. Hence, in this case, M^3 is topologically equivalent to S^3.

If X^3 has one handle, one of $f^{-1}(X^3)$, $S^3 - f^{-1}(\text{Int } X^3)$ is a solid torus. If $S^3 - f^{-1}(\text{Int } X^3)$ is a solid torus, each of X^3, M^3 - Int X^3 is a solid torus and no counterexample results. If $f^{-1}(X^3)$ is a solid torus, there is a homeomorphism h of $f^{-1}(X^3)$ onto X^3 that agrees with f on Bd $f^{-1}(X^3)$ and M^3 is homeomorphic to S^3.

One might try proving the Poincaré conjecture by reducing the number of handles in X^3. If one can find a handle in $f^{-1}(X^3)$ one can cut off this handle and redefine f so that the images of the singularities of the modified f lie in a cube with fewer handles than X^3 has.

Question 7. What special can be said about a cube with handles tamely embedded in S^3 so that some simple closed curve in Bd X bounds a disk in S^3 - Int X but none on Bd X? It might be of help in reducing the handles on X^3 of Theorem 1 if one knew, for example, that such a Bd X contained two transverse simple closed curves one of which bounded a disk in X and the other of which bounded a disk in E^3 - X. This may be difficult to show since I do not believe it is even known to be true in case S^3 - Int X is a cube with handles. As noted in the following paragraph, if we start modifying f and X^3 so as to reduce handles we arrive at a situation where there is a simple closed curve on the modified Bd X^3 which bounds a disk in M^3 - Int X^3 but none on Bd X^3. If Bd X^3 contained two transverse simple closed curves one of which bounded a disk in X^3 and the other a disk D in M^3 - Int X^3, we could remove another handle from X^3 by thickening D and adding it to X^3.

It follows from the loop theorem that there is a disk D in S^3 such that $D \cap f^{-1}(X^3) = \text{Bd } D \cap \text{Bd } f^{-1}(X^3)$ and Bd D does not bound a disk on Bd $f^{-1}(X^3)$. If $D \subset S^3 - f^{-1}(\text{Int } X^3)$, we have the situation mentioned in the preceding question. If $D \subset f^{-1}(X^3)$ and Bd D fails to separate Bd $f^{-1}(X^3)$, then D shows us where to cut off a handle from X^3 and redefine f as suggested in the paragraph following the proof of Theorem 3. Finally, we suppose D separates $f^{-1}(X^3)$. We suppose f is a homeomorphism near D so that we can remove a thickened $f(D)$ from X^3 and be left with two pieces $X_1{}^3$, $X_2{}^3$ so that $X_1{}^3 \cup X_2{}^3$ contains the images of the singularities of f. By continuing splitting we finally arrive at a finite set of X_i^3's whose union contains the images of the singularities of f and the sum of the handles in the X_i^3's is the same as the number of handles in X^3. Then we find a disk D such that $D \cap f^{-1}(\cup X_i^3) = \text{Bd } D \cap \text{Bd } f^{-1}(\cup X_i^3)$ and Bd D does not separate the $\text{Bd } f^{-1}(X_i^3)$ on which it lies. We join the $f^{-1}(X_i^3)$'s by tubes which miss D to obtain a new $f^{-1}(X^3)$ such that the new X^3 has the same number of handles as the old but also, either f(D) enables us to move a handle from X^3 or there is a simple closed curve on Bd X^3 which does not bound a disk on Bd X^3 but does bound one in $M^3 - \text{Int } X^3$.

Instead of reducing the number of handles in X^3, one instead might try to simplify f^{-1} on the handles. If $D_i \times [0,1]$ is a handle of X^3, it is possible to define f so that each $f^{-1}(D_i \times t)$ is a disk with handles such that the set of singularities of f on $f^{-1}(D_i \times t)$ is a finite number of figure eights. By drilling holes in each $f^{-1}(D_i \times [0,1])$ so as to separate the figure eights in the middle level one can change X^3 (by increasing handles) so that the inverse of each handle is the Cartesian product of an arc and a disk-with-a-handle. Hence, we obtain the following result:

THEOREM 4. For each homotopy 3-sphere M^3 there is a map f of S^3 onto M^3 such that the image of the singular set of f is a bouquet of simple closed curves, the inverse of the vertex point of the bouquet is a cube with holes and the inverse of each other point of the bouquet is a figure eight.

REFERENCES

[1] R. H. Bing, "Some aspects of the topology of 3-manifolds related to the Poincaré conjecture," Lectures on Modern Mathematics, John Wiley and Sons (1964), vol. 2, pp. 93-128.

[2] Wolfgang Haken, "On homotopy 3-spheres," Illinois Journal of Mathematics to appear.

[3] E. E. Moise, "Simply connected 3-manifolds," Topology of 3-Manifolds and Related Topics, Prentice Hall (1962), pp. 196-197.

[4] Russel McMillan, these notes.

CONCERNING FAKE CUBES

by

A. C. Connor

A fake cube is a compact, simply connected 3-manifold with boundary whose boundary is a 2-sphere. It has been shown by R. H. Bing [1] that a fake cube K is a cube if it has a triangulation T such that (1) each vertex of T is on Bd K and (2) each 3-simplex of T has at least two edges on Bd K. The principal object of the present paper is to show that condition (2) can be replaced by (2') each 3-simplex of T has at least one edge on Bd K.

A point of a 2-complex K which has no open 2-cell neighborhood in K is called a singular point or a singularity of K; the set of all singular points of K will be denoted by S(K). Two kinds of singularities will be of interest here. A singular point is of the first kind if it has a neighborhood homeomorphic to the product of an arc and a triod and is of the second kind if it is not of the first kind and has a neighborhood homeomorphic to the cone over a set consisting of a circle together with three of its radii. A compact 2-complex is said to be of Type 1 if each of its singularities is of the first kind and to be of Type 2 if each of its singularities is of the first or second kind.

The following theorem, which is used in the proof of the main result, may be of independent interest.

THEOREM. Every simply connected 2-complex of Type 1 has a nontrivial second homology group.

This theorem is proved by showing that the set of singular points of a 2-complex K of Type 1 is the union of a finite number of disjoint simple closed curves each of which separates K, and that if K is any 2-complex of Type 1 which is separated by every simple closed curve in S(K), then (a) K contains a closed 2-manifold without boundary and (b) $H_2(K, Z_2) \neq 0$. A corollary is that no 2-complex of Type 1 is contractible.

Now suppose that K is a fake cube having a triangulation satisfying (1) and (2'). It follows as in [1] that K has a triangulation T which satisfies (1) and (2'). It follows as in [1] that K has a triangulation T which satisfies (1) and (2') as well as (3) each 2-simplex of T which has two edges on Bd K is a subset of Bd K and (4) no 3-simplex of T has exactly two of its 2-dimensional faces on Bd K.

Suppose T contains more than one 3-simplex. Let T^1 be the first barycentric subdivision of T and let N denote the collection of all simplexes of T^1 which have no vertex on Bd K. It is easily verified that N is a 2-complex of Type 1.

Let R denote the second derived neighborhood with respect to T of Bd K and let $K^1 = \overline{K - R}$. Since K^1 is a regular neighborhood of N, K^1 collapses to N and hence, since K is homeomorphic to K^1, K collapses to N. It follows that N is contractible, which is impossible, since N is a 2-complex of Type 1. Hence T contains just one 3-simplex, so K is a cube.

A further result is that if the Poincaré conjecture is false, then there exists a fake cube K which is not a cube and which has a spine N of Type 2 such that S(N) is connected and each component of N - S(N) is an open disk.

REFERENCE

[1] R. H. Bing, "Some Aspects of the Topology of a 3-Manifold Related to the Poincaré Conjecture," Lectures on Modern Math. Vol. 11 edited by T. L. Saaty, John Wiley & Sons, Inc., 1964.

CHAPTER III

ON CERTAIN FIRST COUNTABLE SPACES

by

R. W. Heath

Some first countable spaces with interesting variety of structures are semi-metric spaces, developable spaces and (first countable) M_1-spaces. Each seems to be characterized by having a basis with a certain kind of uniformity, and in fact these seem to be the three basic types of uniformity in the spectrum from first countable to metric spaces.

Definition 1.1 A T_2-space S is semi-metric if there is a distance function (or "semi-metric") d for S such that (i) for each x and y in S, $d(x, y) = d(y, x) \geq 0$ and $d(x, y) = 0$ only if $x = y$, (ii) for $x \in S$ and $M \subset S$, $\inf\{d(x, y): y \in M\} = 0$ if and only if x is in the closure of M.

Definition 1.2 A development for a space S is a sequence G_1, $G_2, \ldots$ of open coverings of S such that, for each $p \in S$ and each open set R containing p, there is an n such that every element of G_n containing p lies in R. A developable space is a T_2-space having a development.

REMARKS (1) For any metric space S, the sequence $G_1, G_2, \ldots$ in which, for each i, G_i is the set of all open sets of diameter less than $1/i$ would be a development for S.

(2) A space X with development $G_1, G_2, \ldots$ has the natural semi-metric d: $d(x, y) = \inf\{1/n: x \in G$ and $y \in g$ for some $g \in G_n\}$ (where we have let $X \in G_1$ and $G_i \supset G_{i+1}$ for all i —without loss of generality). Note that spherical neighborhoods under d will be open, but d is not continuous—hence cannot be a metric.

(3) Regular developable spaces are Moore space, i.e., satisfy the first three parts of Axiom 1 in [32].

(4) Russian mathematicians call semi-metric spaces "symmetrizable" and they call a development a "countable complete family of open coverings", also what they call a space with a "uniform base" is simply a pointwise paracompact developable space and a "symmetrizable Cauchy space" is equivalent to a developable space. This difference in terminology led to the

unfortunate situation that they did not know until 1962 that there was a non-developable semi-metric space (for such a space see [28], [17] or Example 2.1 below). In fact the only "example" they now have, by A. Lunts in 1962 Doklady, is incorrect (the set of all countable ordinals with the order topology!). This communications problem has also led to the recent rediscovery of a number of theorems.

Definition 1.3 A collection A is called closure-preserving, if, for every $B \subset A$, the union of the closures of members of B is closed. A collection is σ-closure-preserving if it is the union of the countably many closure-preserving collections. An M_1-space is a T_3-space with a σ-closure-preserving basis.

REMARK (5) An M_1-space is paracompact by Professor E. Michael's theorem [30]: A T_3-space in which every open cover has a σ-closure-preserving open refinement is paracompact. A semi-metric for a first countable M_1-space can be defined in somewhat the same way as for a developable space.

In Summary we have

first countable M_1 ⟹ paracompact semi-metric

metric ⟹ ⟹ first countable

developable ⟹ semi-metric

and examples show that none of the arrows can be reversed—except that developable M_1-spaces are metric.

2. Examples

Example 2.1 [28]. A paracompact semi-metric space which is an M_1-space but not developable.

The points of S are the points of E^2 and the semi-metric d is defined as follows. If either x or y is on the x-axis, let $d(x, y) = |x-y| + a(x, y)$, otherwise let $d(x, y) = |x-y|$, where $|x-y|$ is the ordinary Euclidean distance and $a(x, y)$ is the (radian) measure of the smallest non-negative angle formed by a line through x and y with a horizontal line. Thus a "spherical" neighborhood of a point on the x-axis with respect to d is "bow-tie" shaped.

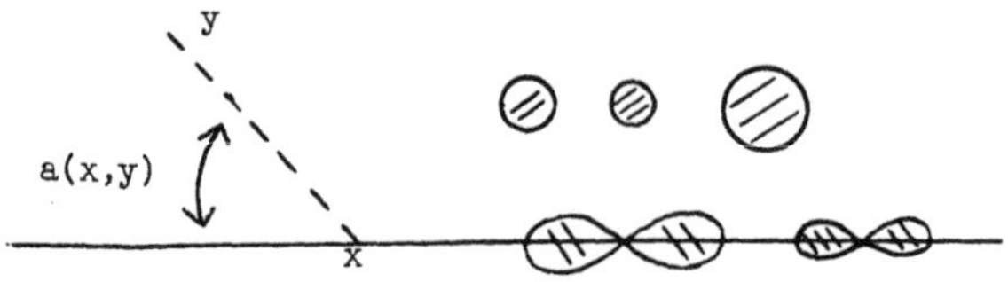

These bow-tie neighborhoods of points on the x-axis and discs elsewhere are used as a basis for the topology.

REMARKS (1). If one uses as basis elements bow-tie regions at every point of E^2 (and let $d(x, y) = |x-y| + a(x, y)$ for all x and y), the space is then a semi-metric space which is not paracompact or an M_1-space. Also, though it is connected, locally connected and complete, it is not arc-wise connected [17].

(2) The semi-metric space S with points those of E^2 and a basis consisting of all bow-tie regions centered on irrational points of the x-axis and open discs centered on all other points is paracompact but has the following property: there is no semi-metric for S with respect to which all spherical neighborhoods are open sets [15].

Examples of developable spaces are given in [25] and [39] of these proceedings.

3. Metrization of Developable Spaces

Professor R. B. Jones proved the following in [22]:

THEOREM 3.1. If $2^{\aleph_0} < 2^{\aleph_1}$ then every separable normal space is $\aleph_1$-compact (i.e., every uncountable subset of it has a limit point).

COROLLARY 3.2. If $2^{\aleph_0} < 2^{\aleph_1}$ then every normal separable Moore space (regular developable space) is metrizable.

Much effort has been made by many people to rid the corollary, in one way or another, of the assumption $2^{\aleph_0} < 2^{\aleph_1}$. There has been very little success (see [5], [20], [35], [39]), although the converse to Theorem 3.1 was proved in [21] so that:

THEOREM 3.2. Every separable normal T_3-space is $\aleph_1$-compact if and only if $2^{\aleph_0} < 2^{\aleph_1}$.

There is also the following result in which the hypothesis $2^{\aleph_0} < 2^{\aleph_1}$ is replaced by another, possibly weaker, hypothesis about the reals (and equivalence to the metrization proposition is shown).

THEOREM 3.3. [20] Every separable normal Moore space is metrizable if and only if every uncountable subset M of E^1 contains a subset which is not a G_δ set (relative to M).

It was also conjectured by Professor Jones in [22] that every normal Moore space is metrizable. The only major results in this direction have been the following due to Professor R. H. Bing in [6].

THEOREM 3.4. A paracompact Moore space is metrizable.

THEOREM 3.5. A collectionwise normal Moore space is metrizable.

REMARK. It is Theorems 3.4 and 3.5 in particular which have been rediscovered in several forms by the Russians (for example see [1, p. 44], [2, p. 40]). (Refer back to Remark 4 in Section 1 for equivalent terminology).

For further details on the normal Moore space metrization problem, see [25] and [39] or [23].

Finally, a proposition which I think might be true (but will probably require the assumption $2^{\aleph_0} < 2^{\aleph_1}$ —because of Theorem 3 in [20]) is: Every normal pointwise paracompact Moore space is metrizable.

4. Metrizability and Developability of Semi-metric Spaces.

There are the following characterizations of developable and metrizable spaces in terms of special semi-metrics possessed by the space.

THEOREM 4.1 [3]. A T_3-space S is developable if and only if there is a semi-metric d for S such that whenever

$$\lim_{n \to \infty} d(x_n, p) = \lim_{n \to \infty} d(y_n, p) = 0$$

then

$$\lim_{n \to \infty} d(y_n, x_n) = 0 .$$

THEOREM 4.2 [37]. A T_3-space S is metrizable if and only if there is a semi-metric d for S such that, whenever

$$\lim_{n \to \infty} d(x_n, p) = \lim_{n \to \infty} d(x_n, y_n) = 0$$

then

$$\lim_{n \to \infty} d(y_n, p) = 0 .$$

A question posed by Morton Brown [8] remains open, however: "What 'topological' property can be added to a semi-metric space to get a Moore space?" The following property (suggested by Professor Jones) is a rather natural candidate: for every compact subset M of the space, there is a sequence $U_1, U_2, \ldots$ of open sets such that, for every open set R containing M, there is an n such that $M \subset U_n \subset R$. Unfortunately, I can give an example of a non-developable semi-metric space (which is paracompact) which does have that property.

Example 4.1. Let M be an uncountable subset of the real numbers all of whose compact subsets are countable ([27, p. 422]). Let S be the space whose points are the points of E^2 not on the x-axis or belonging to a subset M^1 of the x-axis congruent to M, and let a basis consist of all open disks centered on points of $S - M^1$ and of all bow-tie regions (see Example 2.1) centered on points of M^1. Then S is semi-metric but not developable, and it can be shown that S has the above property, that there is a "countable neighborhood system" for each compact subset of S.

Some theorems which might be helpful in answering Morton Brown's question are to be found in [17] and [36]. The following, for example, is a reformulation of Theorem 3.6 of [17].

THEOREM 4.3. A regular semi-metric space S is a Moore space if there is a decreasing sequence $G_1 \supset G_2 \supset \cdots$ of open coverings of S such that, if $p \in S$, then any decreasing sequence, $g_1 \supset g_2 \supset, \cdots$ with $p \in g_i \in G_i$ $(i = 1, 2, \ldots)$ is a local base for p, and for each $p \in S$ there is such a decreasing sequence.

The following two theorems, 4.4 and 4.5, answer Brown's question in certain special cases.

Definition. A semi-metric space S is strongly complete if there is a semi-metric d for S with respect to which every nested sequence $M_1 \supset M_2 \supset \cdots$ of closed sets such that for each n there is some point p_n for which $M_n \subset \{y: d(y, p_n) < 1/n\}$ has non-empty intersection.

THEOREM 4.4 [17]. A strongly complete, regular semi-metric space is a Moore space.

THEOREM 4.5 [19]. A semi-metric space with a point-countable base is developable.

REMARK. A base B is point-countable if each point of the space belongs to only countably many members of B. A pointwise paracompact developable space (same as a T_2-space with a "uniform base" in Russian terminology—see [2]) has a point-countable base. There does exist a Moore space with a point-countable base which is not pointwise paracompact [14]. See [12] for some theorems that can be extended from metric spaces to pointwise paracompact Moore spaces (but not to Moore spaces in general).

The following metrization theorem shows how strong "strong completeness" is.

THEOREM 4.6 [21]. A strongly complete separable semi-metric space is metrizable.

Finally, I have found Theorem 4.7 extremely useful in constructing examples of semi-metric spaces.

THEOREM 4.7 [17]. A T_2-space S is semi-metric if and only if there is a collection $\{g_n(x): x \in S, n = 1, 2, \ldots\}$ of open sets such that (1) for each $x \in S$, $\{g_n(x): n = 1, 2, \ldots\}$ is a local base for x and (2) if $y \in S$ and, for each n, $x_n \in S$ such that $y \in g_n(x_n)$, then $x_1, x_2, \ldots$ converges to y.

REMARK. If in addition $\{g_n(x): x \in S, n = 1, 2, \ldots\}$ satisfies (3), below, then S is developable, and if it satisfies (3) and (4), S is metrizable [17]: (3) If $x \in S$ and y and z are point sequences in S such that, for each n, $x \in g_n(y_n)$ and $z_n \in g_n(y_n)$, then $z_1, z_2, \ldots$ converges to x.

(4) For each x and y in S and each n, $x \in g_n(y)$ implies that $y \in g_n(x)$. Compare these conditions with Theorem 5.1 below.

5. Semi-metric, M_1- and Related Spaces

In [9], Jack Ceder defines M_1-, M_2- and M_3-spaces. M_1-spaces are defined as above in Section 1, as a T_3-space with a σ-closure-preserving basis, and an M_2-space is similarly defined but in terms of a "basis" whose members are not necessarily open. A characterization of (first countable) M_3-spaces (which are defined by Ceder in terms of cushioned collections) is given by Theorem 5.1. By Ceder's definition "M_1 implies M_2" and "M_2 implies M_3" but it is not known whether either converse is true (I believe that it is, at least in the first countable case). Note that M_3-spaces are called stratifiable spaces by Borges in [7] and first countable M_3-spaces are the same as Nagata spaces [9].

THEOREM 5.1 [18]. A T_1-space Y is a Nagata (first countable M_3-space if and only if there is a collection $\{g_n(x): x \in Y, n = 1, 2, \ldots\}$ of open sets such that for each $x \in Y$ (1) $\{g_n(x): n = 1, 2, \ldots\}$ is a local base for x and (2) for every neighborhood R of x, there is an n such that $g_n(x) \cap g_n(y) \neq \emptyset$ implies that $y \in R$.

Theorem 5.1 shows the uniformity which characterizes M_3-space (and hence is at inherent in M_1-spaces). Also, a comparison of Theorem 5.1 with the Moore metrization theorem (or the Alexandroff-Uryshon theorem — see [23] and [1, p. 45]) shows that the relationship of semi-metric spaces to first countable M_3-spaces is exactly analogous to that of developable metric spaces.

Moore's Metrization Theorem: Let S be a T_1-space. Then S is metrizable if there is a sequence $G_1, G_2, \ldots$ of open coverings of S such that $p \in S$, and R is an open set containing p, then there is an n such that $p \in g$, $g \in G_n$, $h \in G_n$ and $g \cap h \neq \emptyset$ imply $h \subset R$.

That analogy between the semi-metric-M_3-space and the developable-metric relations suggests that a paracompact semi-metric space should be an M_3-space. Such a conjecture was made by Jack Ceder in [9] (on the basis of other evidence). In [16], however, an example is given of a paracompact semi-metric space which is not M_3. It remains unknown, then, what condition on a semi-metric space gives an M_3- (or M_1-) space.

6. Continuous Images of Metric Spaces

It was shown by V. I. Ponomarev [34] that a T_1-space is first countable if and only if it is the continuous open image of a metric space. In [4] A. Arhangelskii showed that a T_1-space has a uniform basis (i.e., is a pointwise paracompact developable space) if and only if it is the image of a metric space under a bicompact open mapping ($f: X \to S$ is <u>bicompact</u> if $f^{-1}(p)$ is bicompact for every $p \in S$). This suggested to me the following characterization of developable spaces.

THEOREM 6.1 [18]. A T_2-space S is developable if and only if there is a metric space X and an open mapping f from Y onto S such that for every point $p \in S$ and every open set R containing p there is an $\varepsilon > 0$ such that $f(N(f^{-1}(p), \varepsilon) \subset R$ (where $N(f^{-1}(p), \varepsilon)$ denotes the spherical neighborhood of radius ε about $f^{-1}(p)$, $N(f^{-1}(p), \varepsilon) = \{x: |x-y| < \varepsilon$ for some $y \in f^{-1}(p)\}$).

<u>REMARKS</u>. (1) If in place of "every <u>point</u> $p \in S$" (in the hypothesis of Theorem 6.1) we put "every compact subset $p \subset S$," the conclusion would be that S is metrizable. Similar characterization of semi-metric and first countable M_3-spaces are also given in [18].

(2) Another way of stating Theorem 6.1 would be: "Every developable space is a decomposition space of a metric space corresponding to a lower-semi-continuous decomposition M with the following open sets $N_\varepsilon(g)$ sufficing for a basis: for each $g \in M$ and each $\varepsilon > 0$, let $N_\varepsilon(g) = \{h:$ for some $x \in g$ and $y \in h$ $|x-y| < \varepsilon\}$."

<u>Question</u>: Under what condition on the function f is the space S of Theorem 6.1 normal? An answer to this question might make it possible to use that characterization of developable spaces to make some progress on the normal Moore space metrization problem.

One might restrict one's attention to continuous images of "nice" mappings of general metric spaces. A T_3-space which is the continuous image of a separable metric space is called a <u>cosmic space</u>. Such spaces are clearly Lindelöf, separable and perfectly normal. For some other properties see [31], from which comes the following characterization of cosmic spaces (see Lemma 4.1 of [31]).

THEOREM 6.2. A T_3-space S is a cosmic space if and only if there is a countable collection $\{B_1, B_2, \ldots\}$ of subsets of S such that for each $p \in S$ and each open set R containing p there is an n such that $p \in B_n \subset R$.

A cosmic space then has what one might call a countable pseudobase (except that Professor Michael has already called something else a pseudobase in [31]). The following questions were raised by C. J. R. Borges in [7]: (1) Is every separable M_3-space a cosmic space? (2) Is every cosmic space an M_3-space? and (3) Is every countable T_3-space (which would of course be a cosmic space) an M_3-space? The examples in [16] and [13] give negative answers to (2) and (3) respectively, and it has been shown in [13] that a first countable cosmic space is semi-metric at least, but question (1) remains open and one can ask also:

<u>Question</u>: Is every Lindelöf semi-metric space a cosmic space?

7. Summary

In conclusion some of the known relations among first countable spaces are:

(1) A paracompact Moore space is metrizable.

(2) A separable normal Moore space is metrizable (assuming the continuum hypothesis or some related hypothesis).

(3) There is a characterization of developable spaces as lower-semi-continuous decompositions of metric spaces—i.e., open continuous images of metric spaces—of a certain type.

(4) A semi-metric space can be far from being developable—it may even fail to have a semi-metric under which all spherical neighborhoods are open.

(5) A semi-metric space which is strongly complete or which has a point countable base is developable, but not conversely (even for complete Moore spaces in the first case). There is still lacking a satisfactory "topological" characterization of those semi-metric spaces which are developable.

(6) Developable spaces can be characterized as those having a semi-metric d such that $x_i \to p$ and $y_i \to p$ implies $d(x_i, y_i) \to o$. Metric spaces are those having a semi-metric d such that $x_i \to p$ and $d(y_i, x_i) \to o$ implies $y_i \to p$.

(7) A first countable M_1-space is a paracompact semi-metric space, but the converse is not true; nor is it true that a cosmic space (continuous imate of a separable metric space) is necessarily an M_1-space.

(8) A first countable cosmic space is a Lindelöf semi-metric space.

The following are some questions that remain open.

1. Is every normal Moore space metrizable? Is every pointwise paracompact normal Moore space metrizable (assuming the continuous hypothesis if necessary).

2. Is every normal semi-metric space paracompact? (There may be a counter example to this which is not a counter example to question 1.)

3. What is a necessary and sufficient "topological" condition for a semi-metric space to be developable?

4. What is a necessary and sufficient condition for a semi-metric space to be an M_1-space?

5. Is every M_3-space an M_1-space?

6. Is every separable M_1-space a cosmic space? Is every Lindelöf semi-metric space a cosmic space?

REFERENCES

[1] P. S. Alexandrov, "On some results concerning topological spaces and their continuous mappins," Proc. Sympos. Prague (1961), pp. 41-54.

[2] _____, "Some results in the theory of topological spaces, obtained within the last twenty-five years," Russian Math. Surveys, 15 (1960), pp. 23-83.

[3] P. S. Alexandrov and V. V. Nemitskii, "Der allgemaine Metrisatienssatz und das Symmetricaxiom " (in Russian; German Summary), Mat. Sbornik 3 (45) 1938 pp. 663-672.

[4] A. Arhangel skii, "On mappings of metric spaces," Soviet Math. Dokl. 3 (1962)pp. 953-956.

[5] R. H. Bing, "A translation of the normal Moore space conjecture," Proc. Amer. Math. Soc., 16 (1965) pp. 612-619.

[6] _____, "Metrization of topological space," Canad. J. Math., 3(1951) pp. 175-186.

[7] C. J. R. Borges, "On stratifiable spaces," to appear.

[8] Morton Brown, "Semi-metric spaces," Summer Institute on Set-Theoretic Topology, Madison, Wisconsin, Amer. Math. Soc., (1955) pp. 62-64.

[9] Jack Ceder, "Some generalizations of metric spaces," Pacific J. Math., 11 (1961) pp. 105-125.

[10] D. R. Taylor, "Metrizability in normal Moore spaces," Pacific J. Math., to appear.

[11] B. Fitzpatrick and D. R. Traylor, "Two theorems on metrizability of Moore spaces," Pacific J. Math., to appear.

[12] E. E. Grace and R. W. Heath, "Separability and metrizability in pointwise paracompact Moore spaces," Duke Math. J., 31 (1964) pp. 603-610.

[13] R. W. Heath, "A note on cosmic spaces," to appear.

[14] _____, A non-pointwise paracompact Moore space with a point-countable base," to appear.

[15] _____, "A regular semi-metric space for which there is no semi-metric under which all spheres are open," Proc. Amer. Math. Soc., 12 (1961) pp. 810-811.

[16] _____,"A paracompact semi-metric space which is not an M_3-space," to appear.

[17] _____, "Arcwise connectedness in semi-metric spaces," Pacific J. Math., 12 (1963) pp. 1301-1319.

[18] _____, "On open mappings and certain spaces satisfying the first countability axiom," Fundamenta Mathematicae, to appear.

[19] _____, "On spaces with point-countable bases," Bull. Pol. Akad. Nauk., to appear.

[20] _____, "Screenability, pointwise paracompactness and metrization of Mocre spaces," Canad. J. Math., 16 (1964) pp. 763-770.

[21] _____, "Separability and $\aleph_1$-compactness," Coll. Math., 12 (1964) pp. 11-14.

[22] F. B. Jones, "Concerning normal and completely normal spaces," Bull. Amer. Math. Soc., 43 (1937) pp. 671-677.

[23] _____, "Metrization," American Math. Monthly, to appear.

[24] _____, "Moore spaces and uniform spaces," Proc. Amer. Math. Soc., 9 (1958) pp. 483-486.

[25] _____, "Remarks on the normal Moore space metrization problem," these proceedings.

[26] _____, R. L. Moore's Axiom 1' and metrization," Proc. Amer. Math. Soc., 9 (1958) p. 487.

[27] C. Kuratowski, "Topologie I," 4th ed. Monografie Mathmatyczne, vol. 20, Pánstwowe Wydawnictwe Naukowe, Warsaw, 1958.

[28] L. F. McAuley, "A relation between perfect separability, completeness and normality in semi-metric spaces," Pacific J. Math., 6 (1956) pp. 315-326.

[29] _____, "A note on complete collectionwise normality and paracompactness," Proc. Amer. Math. Soc., 9 (1958) pp. 769-799.

[30] E. A. Michael, "Another note on paracompact spaces," Proc. Amer. Math. Soc., 8 (1957) pp. 822-828.

[31] _____, $\aleph_0$-spaces, to appear.

[32] R. L. Moore, "Foundation of Point Set Theory," Am. Math. Soc. Coll. Publ. 13, Revised Edition, (Providence, 1962).

[33] V. W. Niemytzki, "On the 'third axiom of metric spaces'," Transactions Amer. Math. Soc., 29 (1927) pp. 507-513.

[34] V. I. Ponomarev, "Axioms of countability and continuous mappings," Bull. Pol. Akad. Nauk., 8 (1960) pp. 127-134.

[35] D. R. Traylor, "Normal separable Moore spaces and normal Moore spaces," Duke Math. J., 30 (1963) pp. 485-493.

[36] J. M. Worrell, Jr., and H. H. Wicke, "Characterizations of developable topological spaces," Canad. J. Math., 17 (1965) pp. 820-830.

[37] W. A. Wilson, "On semi-metric spaces," Amer. J. Math., 53 (1931) pp. 361-373.

[38] J. N. Younglove, "Two conjectures in point set topology," these proceedings.

[39] ____, "Concerning metric subspaces of non-metric spaces," Fund. Math., 48 (1959) pp. 15-25.

REMARKS ON THE NORMAL MOORE SPACE METRIZATION PROBLEM

by

F. Burton Jones

As far as I know the first example of a non-metric Moore space was discovered (probably by Moore himself) in the late 1920's. It is a dendron D in the plane with uncountably many endpoints which has the relative topology of the plane at all of its points except at the endpoints and may be roughly described as follows:

Let C be the standard (ternary) Cantor set on $[0, 1]$ of the x-axis. Designate the components of $[0, 1] - C$ (in order of length and from left to right for those of the same length) by I_1; I_{11}, I_{12}; I_{111}, I_{112}, I_{122}; ... . Select a point p_1 one unit below the midpoint of I_1 and from p_1 draw two straight line (closed) intervals one half-way toward the center of I_{11} and another half-way toward the center of I_{12}. Let p_{11} and p_{12} denote the non-p_1 endpoints of each of these arcs respectively. Repeat the process by drawing two intervals from p_{11} half-way toward the centers of I_{111} and I_{112} and two intervals from p_{12} half-way toward the centers of I_{121} and I_{122}. Continue this process indefinitely and the union of C with all of the arcs so constructed is the dendron D. From p_1 to a point c of C there is only one arc T(c). Any half-open subarc (z, c] of T(c) plus appropriate half-open arcs at each branch point (these are not to contain any other branch point) is to be an open set containing c. Under this change of topology, no point of C is a limit point of C in the space D.

Moore had another example derived from this which he called the automobile road space which went something like this:

Start two roads out from the origin in opposite directions. After each has gone a mile let each branch into two roads. Continue each of the now four roads for a mile and let each branch into two roads. Continue this a countable infinity of times, so that none of the new branches ever intersect and so that all roads (regardless of which turns are taken) proceed indefinitely far from the origin. Now, when each road (there are c of them) comes to the edge of the plane, continue it straight on away from the plane another countable infinity of miles. The boundaries of the roads do not belong to the space. Neighborhoods (or regions) are now open disks and quite a number of Moore's other axioms hold true in the autom bile road space. In particular, Axioms 0, 1, 2, 3, 4, 5, 5_1 and 5_2[1] hold true.

The usual way to see that these spaces were not metric was to observe that each contained a closed separable subspace (the whole space in the case of D) which was not perfectly separable (i.e., the subspace had no countable topological basis). This kind of observation left me mildly restless and it was only after I discovered that neither was normal that I felt that I was closer to the "real reason" for non-metrizability.

The argument I used to prove the non-normality of D (and the automobile road space contains D as a subspace) has a good deal in common with Professor Younglove's and goes like this:

Let M denote the set of all points m of C different from 0 and 1 such that m is not a limit point of a component of [0, 1] - C. In the space D both M and C - M are closed. If the space D were normal, there would exist for each m in M, a half-open arc (z, m] of T(m) [T(m) was that unique arc from p_1 to m] such that in the space D, $\bigcup_{m \in M}$ (z, m] has no limit point in C - M. Now let f be a function from [0, 1] to the non-negative real numbers defined as follows: f(x) = |img (z)| where x belongs to M; otherwise, f(x) = 0. Since the set of points of discontinuity of f is exactly the set M, M (as a subset of [0, 1] with its usual topology) must be an F_σ. This is obviously impossible since it would imply that [0, 1] is of the first category.

In fact, a little expansion of this argument was the argument I first got to prove that if $2^{\aleph_0} < 2^{\aleph_1}$ then every separable normal Moore space is metric. For suppose that S is a separable normal Moore space with a monotone descending development $G_1, G_2, G_3, \ldots$. Let $p_1, p_2, p_3, \ldots$ be a countable dense subset of S and suppose that M is an uncountable subset of S - $\{p_1 + p_2 + p_3 + \cdots\}$ which has no limit point (if no such set M exists, S is perfectly separable and hence metric and normal). For each point m of M let $m_1, m_2, m_3, \ldots$ be the sequence of integers such that for each i, m_i is the smallest natural number n having the property that p_n belongs to an element of G_i which contains m. Evidently $p_{m_i} \to m$. Using these sequences, $\{m_i\}$, the points of M may be given the lexicographic order and embedded in an order preserving manner in [0, 1] of the real numbers. Let T denote a 1-1 transformation that thus embeds M into [0, 1].

Without loss of generality we may assume that every point of T(M) is a point condensation of T(M). Since M is uncountable (but of cardinality $\leq$ c) and the number of countable subsets of M is c (assuming that $2^{\aleph_0} < 2^{\aleph_1}$) there are more than c subsets H of M such that each of T(H) and T(M-H) is dense in the other.

Now for each such set H, let f be a real valued (non-negative) function f from T(M) defined as follows:

Since S is normal there exists an open set U such that $M - H \subset U \subset \overline{U} \subset H$. If k is a point of $M - H$, let n be the smallest natural number such that every element of G_n containing k is a subset of U and let $f[T(k)] = 1/n$; otherwise, let f be zero (i.e., $f(T(H)) = 0$). On T(M), f is continuous exactly at points of T(H) (and discontinuous at points of $T(M - H)$); hence T(H) is G_δ-set relative to T(M). So there exists a sequence Q_1, Q_2, Q_3, of open subsets of [0, 1] such that $(Q_1 \cdot Q_2 \cdot Q_3 \cdot \cdots) \cdot T(M) = T(H)$. Now let $Q(H) = Q_1 \cdot Q_2 \cdot Q_3 \cdot \cdots$ Then Q(H) is a G_δ-set in [0,1]. But suppose that H' is another subset of M of the type that H is and that Q(H') is the corresponding G_δ-set in [0, 1]. If H and H' are different then so are Q(H) and Q(H') different. Since there are only c distinct G_δ-sets in [0, 1] we have a contradiction. So no such set M exists and S is metric and normal.

Among several examples of non-separable non-metric Moore space there is one which I had high hopes of proving to be normal. (I don't believe that I ever proved it was not normal.) It is a tree branching upward as did space D but continuing through uncountably many branching levels (in a well-ordered fashion) such that each branch at each branching level continues upward along a countable infinity of branches instead of two. It can be started in the plane but must be continued uncountably many times out beyond the plane producing an uncountable monotone increasing well-ordered sequence of connected domains whose union is the space.

Start with a point p_0 and from p_0 draw closed straight line intervals of length 1/2, 1/4, 1/8,... which have only p_0 in common. Call the set of non-p_0 endpoints of these intervals M_1. Repeat this process at each point of M_1 producing a new level of end points which we shall call M_2. While M_2 is a countable infinity of infinities it is still countable. Now continue this process to produce an uncountable sequence of branching levels $M_0, M_1, M_2, M_3, \ldots, M_z, \ldots$ $(z < \omega_1)$ so that each level is countable and if p belongs to M_z, z' is an ordinal larger than z, and n is a positive integer, there is a point p' belonging to $M_{z'}$ such that there is a path of intervals: pp_1 from p of M_z to p_1 of M_{z+1}, p_1p_2 from p_1 to p_2 of $M_{z+2}, \ldots$ which starts at p and runs consecutively through the intervening branching levels and terminates at p' of $M_{z'}$, and whose length is $1/2^n$. To see how the construction can be made to carry on this property inductively consider the following two cases:

Case 1. Let z' be a non-limit ordinal such that the branching levels have already been defined so as to possess the property and let us see how to define $M_{z'}$. We simply go from $M_{z'-1}$ to $M_{z'}$ exactly as we did from M_1 to M_2. From a point p at any lower level M_z $(z < z'-1)$ there would be a path to q in $M_{z'-1}$ of length $1/2^{n+1}$. Join this path the one from q to a point p' of $M_{z'}$ of length $1/2^{n+1}$ and one gets the required path of length $1/2^n$ from p of M_z to p' of $M_{z'}$.

Case 2. Let z' be a limit ordinal less that ω_1 such that the inductive property is possessed for all smaller ordinals. Let z be an ordinal less than z and let $z < z_1 < z_2 < \ldots$ be a simple countable monotone increasing sequence of ordinals such that $z_i \to z'$. From p in M_z to a point p_1 in M_{z_1} there is a path of length $1/2^{n+1}$, from p_1 to a point p_2 in M_{z_2} there is a path of length $1/2^{n+2}$, etc. Now, terminate this path in a point p' which we shall put in $M_{z'}$. The path will have length $1/2^n$ and since this is done countably many times (once for each n) for p and there are only countably many such points p (when z' is countable), the process can be continued uncountably many times carrying on the inductive property.

Obviously, no path intersects all branching levels for if it did there would be some natural number n such that its length between consecutive levels would be $1/2^n$ uncountably many times and hence after the first infinity it would be too long to have been continued because all paths have finite length, it being possible to back up in a well-ordered sequence only finitely many times.

Now let the space S be the union of all these uncountably many intervals with "region" being one of three types:

1. If a point p is of order 2 let a region be an open (connected) interval of points of order 2 containing it.

2. If a point p is a point of order ω in a non-limit branching level, let a region be the union of all of the half-open intervals that start from (and include) p and go $1/2^n$ of the distance toward the next branch point in whatever direction one has chosen (either up or down).

3. If p is a point of order ω in a branching level M_z where z is a limit ordinal, p got into M_z by being the terminal endpoint of a path P starting from below. Let (x, p] be a half-open subarc of P. Add to (o, p] those half-open intervals which start at p and go up $1/2^n$ of the distance toward the next branch point (as in (2) above), and to this add all regions of type (2) defined for this n whose center (or branch) point belongs to (o, p].

The reader should have no difficulty seeing what the topology of this space is when using the set of all regions as a topological basis. And it should be fairly easy to see that S is a Moore space which is connected, locally connected, locally separable, but not perfectly separable. Such a space cannot be metric.

My original thoughts (perhaps falacious) on normality went something like this:

The space S corresponds to that part of the space D below the Cantor set. Each path in D - C which has p_1 as one of its endpoints and is not a proper subset of another such path is a half-open arc with p_1 as one of its endpoints; the other endpoint is in C. Likewise each path in S which has p_0 as one of its endpoints and is not a proper subset of another such path is a half open arc with p_0 as one of its endpoints. So if one adds to each such half-open arc another point to become an endpoint of that half-open arc, one obtains a new space $\tilde{S}$.

Denote $\tilde{S}$ - S by K. Then the topology of $\tilde{S}$ at points of K is completed just as it was for points of C in the space D. Furthermore, the set K has a natural linear (total) order as did the points of C (lexicographic order of the numbers n, which give the length of the straight intervals as $1/2^n$ forming the path).

Again K is not the first category in the open interval topology of the linear ordering. Furthermore, K is a discrete set in $\tilde{S}$. So if $\tilde{S}$ were normal and H were a subset of K such that H and K - H are dense in each other in the linear order topology, the covering of H and K - H by disjoint open sets U(H) and U(K-H) would give rise to two real functions f and g each continuous at each point of K where the other is discontinuous. Again this would force K to be of the first category. Hence $\tilde{S}$ is not normal.

However, in the analogy D - C was metric (and hence normal). So perhaps $\tilde{S}$ - K, that is, S is normal. I think I was never able to settle this question. I do recall proving that K with its linear order topology was not a Souslin space (it is not separable but it does contain an uncountable collection of disjoint open sets) but this didn't seem to help.

REFERENCES

[1] R. L. Moore, Foundations of Point-set Theory, A. M. S. Colloquium Publication, vol. 13 (1932).

[2] F. B. Jones, "On certain well-ordered monotone collections of sets," Journal of the Elisha Mitchell Scientific Society, vol. 69 (1953) pp. 30-34.

TWO CONJECTURES IN POINT SET THEORY

by

J. N. Younglove

Two conjectures in topology are discussed in this paper. One, referred to as the normal Moore space conjecture due to F. B. Jones [4] states that every normal Moore space (i.e., satisfies Axiom 0 and the first three parts of Axiom 1 of [9]) is metrizable. The other due to C. H. Dowker [7] conjectures that every countably paracompact Hausdorff space is normal. An example of R. H. Sorgenfrey [6] is shown to be an example of a paracompact space S such that S^2 is not countably paracompact.

Traylor [1] has shown that every locally compact, normal, non-metrizable Moore space contains an uncountable discrete collection of points with respect to which the space is not collectionwise normal. That is, there is no collection of disjoint open sets such that each open set of the collection intersects the set in exactly one point. Bing [2] has shown that each separable, non-metrizable Moore space contains such an uncountable set. The following is a description of a class of separable Moore spaces which is of considerable interest.

Let H denote a subset of the x-axis in the plane. Let Σ denote the Moore space whose points are the points of the upper half-plane together with the points of H and having the development $\{G_i\}_1^\infty$ such that for each integer n, G_n is the collection of sets which are either (a) interiors of circles of radius less than $1/n$ lying wholly above the x-axis or (b) the interiors of such circles tangent to the x-axis from above at a point H together with the point of tangency. The space Σ will be said to be derived from H.

The set H is discrete in Σ and the metrizability of Σ is equivalent to asserting that H is a countable set. Thus, if there exists an uncountable subset H of the x-axis such that the derived Moore space is normal, then Jones' question [4, p. 676] would be answered. Jones [4] has shown that a set of cardinality of the continuum in a normal separable Moore space must contain a limit point of itself and, consequently, any set H which gives rise to a counterexample as described must have cardinality less than C.

Heath [8] has shown that the existence of a separable, normal, non-metrizable Moore space implies the existence of an uncountable subset H of the x-axis such that every subset of H is the intersection of H and a G_δ set in the reals. Bing [3] has shown that if H is such a set, then the derived Moore space is a separable, normal, non-metrizable Moore space. By using a theorem of Mazurkiewicz [5, p. 236] which states that a G_δ set in the reals can be expressed as the union of two sets, one of which is empty or a homeomorphic image of the irrationals and the other is at most countable, we see that every uncountable G_δ set in the reals has cardinality C. Thus, to achieve a counterexample, the set H can contain only countable G_δ sets.

In the search for a counterexample to Dowker's conjecture, the above considerations assure the non-normality of a Moore space Σ derived from an uncountable G_δ set on the x-axis. The question then arose as to whether any such Moore space could be countably paracompact and the following result was established.

THEOREM. Suppose H is a G_δ set in the x-axis and Σ is the Moore space derived from H. If Σ is countably paracompact, then H is countable.

<u>Proof</u>: Let H denote an uncountable G_δ set in the x-axis. Let $E_1, E_2, E_3, \ldots$ be an infinite sequence of open subsets of the x-axis whose intersection is H. Using Mazurkiewicz's theorem, there is a subset K of H which is homeomorphic with the irrationals. This allows us to assert the existence of a countable subset A of H such that if x is in A and u is an open interval of the x-axis containing x, then $u \cap H$ is uncountable. Let $x_1, x_2, x_3, \ldots$ be a sequential ordering of the points of A. Let G be a countable open covering of S, the set of points of Σ such that if g is an element of G, then g is S - A or g is an element of G_1 which contains some point of A. We may assume for convenience that no two sets of G which contain points of A intersect. Let W be a refinement of G. Let D_1 denote the union of the elements of W which intersect A and D_2 denote the union of the remainder of the sets in W. There is a region g_1 of G_1 containing x_1 such that (i) $g_1 \subseteq D_1$ and (ii) the closure of the vertical projection u_1 of g_1 into the x-axis is a subset of E_1. There is an open interval v_1 of the x-axis containing x_1 such that $(u_1 - v_1) \cap H$ is uncountable. Thus, $(u_1 - \bar{v}_1) \cap H$ must contain a point of A. Let x_2' be the first point of A in the order $x_1, x_2, x_3, \ldots$ in this set. Let g_2 be an element of G_1 containing x_2' such that the closure of its projection, $\bar{u}_2$, is a subset of $(u_1 - \bar{v}_1) \cap E_2$. As before, there is an open interval v_2 containing x_2' such that $(u_2 - v_2) \cap H$ is uncountable. Let x_3' denote the first point of A in the given order which is contained in the set $(u_2 - \bar{v}_2)$. This process yields an infinite monotone sequence $\bar{u}_1, \bar{u}_2, \bar{u}_3, \ldots$ of closed intervals

with a point p in their intersection. Since $\bar{u}_n \subseteq E_n$, the point p must be a point of H. Due to our method of selecting the sequence $x_1', x_2', x_3', \ldots$ p cannot be a point of A. Thus, p is a point of H - A.

Let g be a region of G_1 which contains p and is a subset of D_2. The region g must intersect infinitely many regions of the sequence $g_1, g_2, g_3, \ldots$ and this contradicts the fact that Σ is countably paracompact.

In Sorgenfrey's example, S is the real number system with the half-open interval topology. The interval in the plane joining the points (0, 1) and (1, 0) becomes a discrete set in the S^2 topology. By using a countable dense (in the plane topology sense) subset of this interval as we used the set A in the proof of our theorem, we conclude that S^2 is not countably paracompact.

QUESTIONS

1. If a Moore space Σ is a collectionwise normal with respect to each discrete collection of degenerate point sets, is it collectionwise normal?

2. The same as 1 but for a normal Moore space.

3. Is there a subset H of the x-axis such that the Moore space Σ derived from H is normal and non-metrizable and further such that H is a subset of a Cantor set in the x-axis?

REFERENCES

1. D. R. Traylor,"Normal separable Moore spaces and normal Moore spaces," Duke Math. J., 30 (1963) pp. 485-493.

2. R. H. Bing, "A translation of the normal Moore space conjecture," Proc. Amer. Math. Soc., 16 (1965) pp. 612-619.

3. R. H. Bing,"Metrization of topological spaces," Canad. J. Math., 8 (1951) pp. 653-663.

4. F. B. Jones,"Concerning normal spaces," Bull. Amer. Math. Soc., 43 (1937) pp. 671-677.

5. W. Sierpinski, General Topology, 2nd ed., U. of Toronto Press, 1956.

6. R. H. Sorgenfrey, "On the topological product of paracompact spaces," Bull. Amer. Math. Soc., 53 (1947) pp. 631-632.

7. C. H. Dowker, "On countably paracompact spaces," Canad. J. Math., 3 (1951) pp. 219-244.

8. R. W. Heath, "Screenability, pointwise paracompactness and metrization of Moore spaces," Canad. J. Math., 16 (1964) pp. 763-770.

9. R. L. Moore, "Foundations of point set theory," Amer. Math. Soc., Colloq. Pub. vol. 13, rev. ed. 1962.

ALMOST CONTINUOUS FUNCTIONS AND FUNCTIONS OF BAIRE CLASS 1.

by

E. S. Thomas, Jr.

The work presented here grew out of a problem about almost continuous functions (question 2 below). Working together F, Burton Jones and I solved that problem and in the process turned up some ideas which proved useful in investigating functions of Baire class 1. Later on I'll try to give a glimpse of why the study of one class of functions led to the study of the other.

A function f on a topological space X to I (the unit interval) is said to be _of Baire class_ 1 provided $f^{-1}(F)$ is a G_δ in X whenever F is closed in I.

The following fundamental result has been known since around 1907 (cf. [1]; p. 280 et seq.): _If X is separable metric then f is of Baire class 1 if and only if there is a sequence of continuous functions on X to I converging pointwise to f on X._

We are interested in finding topological conditions on the graph of a function f which are necessary and/or sufficient for f to be of Baire class 1.

As will be seen later (Theorem 2) the graph of such a function is a G_δ set in $X \times I$ (if X is separable metric this follows easily from the preceding characterization), however, it is easy to manufacture functions not of a Baire class 1 whose graphs are G_δ's. For example, let $f: I \to I$ be defined as follows: If x is diadic rational in $(0, 1)$ write $x = j/2^k$ in lowest terms and put $f(x) = 1/2 + 1/2^k$. Let f be identically 0 elsewhere in I. The desired properties are easily verified.

This example, and most of the examples one is liable to come up with in a hurry, works because the graph is disconnected. This common misbehavior leads to

Question 1. If f is a function from I to I whose graph is a connected G_δ in $I \times I$, is f of Baire class 1?

A function f from a space X to a space Y is _almost continuous_ provided that if U is an open subset of $X \times Y$ containing the graph of f then U contains the graph of a continuous function.

<u>Question</u> 2. If f is as in question 1, is f almost continuous?

To see that the two questions are related, suppose f is almost continuous and its graph G is a connected G_δ. Corresponding to any sequence $U_1, U_2, \ldots$ of open sets whose intersection is G there is a sequence of continuous functions $f_1, f_2, \ldots$ such that, for each n, the graph of f_n lies in U_n.

Although $\cap_{n+1}^{\infty} U_n = G$ does not a priori imply that the corresponding f_n converge to f pointwise, it is plausible that using connectedness of G we can manufacture a sequence of U_n's which are nice enough (in some way) to yield this convergence.

We remark that a special case of a conjecture of Stallings [2] is that the answer to question 2 is yes, even without assuming the graph is a G_δ. J. Cornette, in an as yet unpublished paper, has answered Stallings' conjecture in the negative.

We shall answer both questions in the negative by means of an example. The construction of the example is not hard but more easily understood with a little motivation.

Let f be a function from I to I whose graph G is a connected G_δ. Easy arguments establish the following facts about $\overline{G}$ (the closure of G):

(1) $\overline{G}$ is a nowhere dense subcontinuum of $I \times I$.

(2) There is a dense G_δ set L in I such that, for each x in L, $\overline{G} \cap \ell_x$ (ℓ_x is the vertical line segment $\{x\} \times I$) is a single point and $\overline{G}$ is locally connected at this point.

(3) $\overline{G}$ is almost irreducible from ℓ_0 to ℓ_1 in the sense that there is a subcontinuum K of $\overline{G}$ which is irreducible from ℓ_0 to ℓ_1 and $\overline{G} - K$ is the union of vertical line segments each of which lies over a point of $I - L$.

In our example, L will be the complement of a Cantor set and $\overline{G}$ will be irreducible.

To begin with we construct a Cantor set C in $I \times I$ which intersects the graph of every continuous function from I into I and such that $C \cap \ell_x$ consists of two points or one point according as x is or is not a diadic rational in $(0, 1)$. C is obtained as the intersection of sets C_1, $C_2, \ldots$. Figures 1 and 2 show how C_2 is obtained from C_1.

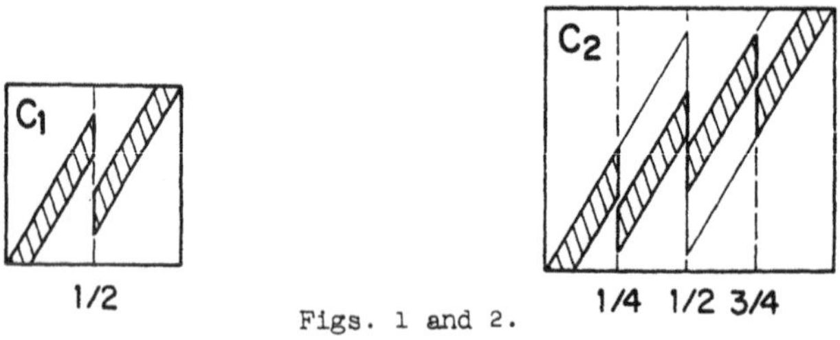

Figs. 1 and 2.

This process is repeated in the obvious way to get C_3, $C_4, \ldots$; the desired properties of C are immediate.

We next construct a function f whose graph is a connected G_δ lying in $(I \times I) - C$. Let K denote the Cantor ternary set in I, let I_1, $I_2, \ldots$ be the collection of complementary intervals of K in I (no interval listed twice), and for each n write $I_n = (r_n, s_n)$.

For a fixed n, f is defined over $[r_n, s_n]$ by specifying its graph as follows. Let z_n and w_n denote the points $C \cap \ell_{r_n}$ and $C \cap \ell_{s_n}$ respectively. Choose a point z_n' on ℓ_{r_n} which lies strictly below and within $1/n$ of z_n and choose a point w_n' on ℓ_{s_n} which lies strictly above and within $1/n$ of w_n. For the graph of f over $[r_n, s_n]$ take any simple arc (simple ≡ intersects each vertical line in $r \to [r_n, s_n]$ exactly once) in $[r_n, s_n] \times I$ running from z_n' to w_n', missing C, and meeting the horizontal lines through 0 and 1. (It is possible to miss C since there is a diadic rational between r_n and s_n). Figure 3 illustrates the process.

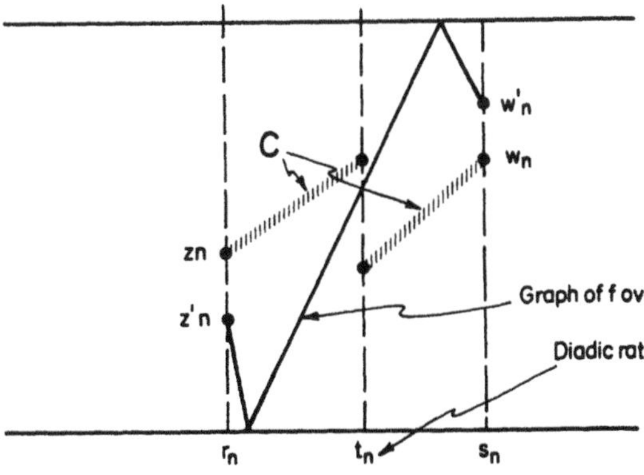

Figure 3

For x in $I - \bigcup_{n=1}^{\infty} [r_n, s_n]$ put $f(x) = 1$ if $x < 1/2$ and $f(x) = 0$ if $x > 1/2$. This completes the definition of f.

Certainly the graph of f is connected and misses C. The fact that it is a G_δ in $I \times I$ follows easily once it is seen that the z_n' and w_n' were chosen to form a countable discrete set, hence a G_δ, in $I \times I$.

Since C meets the graph of every continuous function, f is not almost continuous. To see that f is not of Baire class 1, let e be a point of $[0, 1)$ such that for $x < 1/2$, the set $C \cap \ell_x$ lies strictly below the horizontal line through e. If the set $f^{-1}([0, e])$ were a G_δ in I, then its intersection with $K \cap [0, \frac{1}{2}]$ would also be a G_δ. But this intersection is countable and dense in $K \cap [0, \frac{1}{2}]$, hence cannot be a G_δ.

We next state without proof some positive results.

The first theorem is a specialized result telling what extra condition is needed to get a yes answer to question 1.

THEOREM 1. If the graph G of $f: I \to I$ is connected, then f is of Baire class 1 if and only if $G = \bigcap_{i=1}^{\infty} U_i$ where each U_i is open and simply connected in $I \times I$.

(Theorem 1 fails if connectedness of G is deleted.)

To generalize this result we use the following concept. If X is a topological space and G is a subset of $X \times I$ we say G has property B in case $G = \bigcap_{i=1}^{\infty} U_i$ where each U_i is open in $X \times I$ and, for each i and each x in X, $U_i \cap \ell_x$ is connected.

THEOREM 2. If X is a Hausdorff space such that $X \times I$ is normal and if f is a function on X to I whose graph has property B, then f is the pointwise limit of a sequence of continuous function on X to I.

COROLLARY. If X is separable metric then $f: X \to I$ is of Baire class 1 if and only if its graph has property B.

Question. In order that Theorem 1 go through for functions mapping an n-cube into I, what geometrical condition on the U_i should replace simple connectedness?

REFERENCES

[1] C. Kuratowski, Topologie I, Warsaw (1952).

[2] J. Stallings, "Fixed point theorems for connectivity maps," Fund. Math. 47(1959), p. 249-263.

CHAINABLE CONTINUA

by

J. B. Fugate

A. DEFINITIONS AND ELEMENTARY PROPERTIES

A <u>chain</u> in a topological space is a finite collection $\{E_1, E_2, \ldots, E_n\}$ of open sets such that $E_i \cap E_j \neq \emptyset$ iff $|i-j| \leq 1$. The members of the collection are called <u>links</u>. A chain in a metric space is an ε-<u>chain</u> provided each link has diameter less than ε. A compact metric continuum M is <u>chainable</u> (also called <u>snakelike</u> or <u>arc-like</u>) provided that for each positive number ε, there is an ε-chain covering M. Note that there is no requirement that the links be connected. Examples of chainable continua are: an arc, the pseudo-arc, $Cl\{(x, \sin 1/x): 0 < x < 1\}$, Cl denoting closure.

<u>Proposition</u> 1. Each chainable compact metric continuum M is unicoherent (i.e., if M is the union of two subcontinua A and B of M, then $A \cap B$ is a continuum). A circle is not chainable, since it is not unicoherent.

<u>Proposition</u> 2. A chainable compact metric continuum is not a triod. (A continuum M is a <u>triod</u> provided M is the union of three proper subcontinua of M such that the common part of any pair is the common part of all three and is a proper subcontinuum of each.) A letter "Y" (the union of three arcs, having exactly one end point in common) is a triod, hence is not chainable.

(Propositions 1 and 2 are not hard to establish. The basic lemma needed is the following: If M is a continuum and $\{E_1, \ldots, E_n\}$ is a chain covering M such that M intersects E_1 and E_n, then M intersects each link of $\{E_1, \ldots, E_m\}$.)

<u>Proposition</u> 3. Each subcontinuum of a chainable continuum is chainable.

It follows from Proposition 1 that each subcontinuum of M is unicoherent (i.e., M is <u>hereditarily unicoherent</u>); from Proposition 2, it follows that M contains no triod (i.e., M is <u>a-trodic</u>).

B. EMBEDDING CHAINABLE CONTINUA

Proposition 4. Each chainable compact metric continuum M can be embedded in the plane: i.e., there is a homeomorphism $h: M \to E^2$.

In [4], Bing shows that M can be embedded so that h[M] does not separate the plane. Since the property of separating the plane is a topological invariant, it follows that under any embedding g, g[M] does not separate the plane. Indeed, M can be embedded so that h[M] can be chained with connected links. However, this last property demands upon the embedding, and not upon the continuum M.

Bennett has announded [1]:

Proposition 5. If each of M and N is a chainable compact metric continuum, then $M \times N$ can be embedded in E^3.

C. MAPS ON CHAINABLE CONTINUA

Proposition 6. A chainable compact metric continuum M has the fixed-point property [6]; i.e., if $f: M \to M$ is a continuous function, then there is a point $x \in M$ such that $f(x) = x$.

(One can prove this by an easy argument using chains, once one establishes that if f has no fixed point, then there is a positive number d such that each point of M is moved more than d by f.)

Since chainable compact plane continua do not separate the plane, Proposition 6 provides a partial answer to the following:

Question 1. Suppose K is a compact plane continuum which does not separate the plane. Does K have the fixed point property ?

The name "arc-like" comes from the following characterization of chainability:

Proposition 7. A compact metric continuum M is chainable iff for each positive number ε, there is a continuous function $f: M \to [0, 1]$ such that each point in [0, 1] has inverse image of diameter less than ε.

If one adds the requirement that f be monotone, then this forces M to be locally connected, and hence an arc.

Proposition 8. Each chainable compact metric continuum is the continuous image of the pseudo-arc [7, 8].

Even without this result, it follows from Proposition 7 that an arc is the continuous image of the pseudo-arc, thus, so is any locally connected compact metric continuum. Clearly, then, the pseudo-arc can be mapped onto many continua which are not chainable. A characterization of those continua which are continuous images of the pseudo-arc can be found in [7].

D. AN APPLICATION OF INVERSE LIMITS

A useful characterization of chainable compact metric continua is the following:

Proposition 9. A compact metric continuum M is chainable iff there is an inverse limit system $I \xleftarrow{f_1} I \xleftarrow{f_2} I \leftarrow \cdots$, where, for each positive integer i, f_i is a continuous function of the unit interval onto itself, and M is homeomorphic to the inverse limit set. For example, if each bonding map f_i is the same and is the "rooftop function" f, then the inverse limit is an indecomposable chainable continuum.

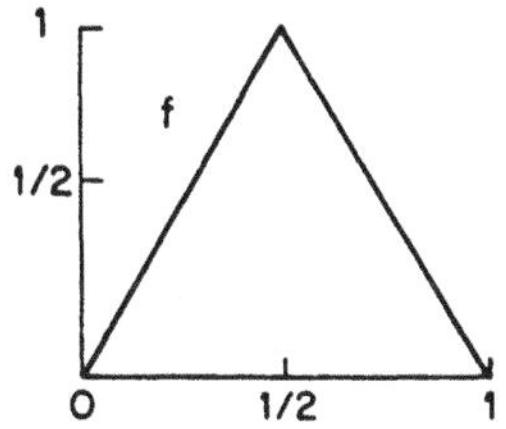

In [9], R. M. Schori utilizes Proposition 9 to construct a chainable continuum U which is universal in the following sense:

Proposition 10. If M is a chainable compact metric continuum, then M can be imbedded in U. U is called a universal snakelike continuum.

N.B. Because of Proposition 8, the pseudo-arc has also been called "universal." Clearly, the continuum U above is not the pseudo-arc, since U contains arcs, while the pseudo-arc is hereditarily indecomposable. Schori has indicated how U can be modified to obtain a chainable compact metric continuum U' with the following properties:

(i) If M is a chainable compact metric continuum, then M can be embedded in U'.

(ii) If M is a chainable compact metric continuum, then M is the continuous image of U'.

E. CONDITIONS IMPLYING CHAINABILITY

The preceding sections should demonstrate that if one knows that a compact metric continuum is chainable, then a great deal can be said about it. In this section, we consider the converse: What are sufficient conditions that a compact metric continuum be chainable? The first answer to this question is due to Bing [4].

Proposition 11. If M is an hereditarily decomposable compact metric continuum, then M is chainable iff M is a-triodic and hereditarily unicoherent.

Another answer to the above question was announced in [5]:

<u>Proposition</u> 12. If M is a compact metric continuum which is the union of two chainable continua, then M is chainable iff M is a-triodic and hereditarily unicoherent.

In considering how one might attempt to extend Proposition 11 to continua which are not hereditarily decomposable, it is informative to consider a continuum which is "nearly chainable." This continuum is the dyadic solenoid, S. S has the following properties:

(i) S is indecomposable, hence it is (trivially) unicoherent and not a triod.

(ii) S is circularly chainable, thus each proper subcontinuum of S is chainable. Hence S is a-triodic and hereditarily unicoherent.

(iii) S is not chainable. Indeed S cannot be embedded in the plane.

In the following result, due to the writer, one eliminates such continua by hypotheses.

<u>Proposition</u> 13. If M is a compact metric continuum such that each indecomposable subcontinuum of M is chainable, then M is chainable iff M is a-triodic and hereditarily unicoherent. This extends Proposition 11, and yielc as a corollary, an extension of Proposition 12.

<u>Corollary</u>. If M is a compact metric continuum which is the union of countably many chainable continua, then M is chainable iff M is a-triodic and hereditarily unicoherent.

F. THE HOMOGENEITY PROBLEM

Still unsettled is the following:

<u>Question</u> 2. What are the homogeneous bounded plane continua?

Bing has shown [2, 3]:

<u>Proposition</u> 14. If M is a chainable compact plane continuum and M is either homogeneous or hereditarily indecomposable, then M is the pseudo-ar(

<u>Question</u> 3. Is the pseudo-arc the only homogeneous compact plane continuum which does not separate the plane?

<u>Question</u> 4. If M is a homogeneous compact plane continuum, is M chainable or circularly chainable ?

<u>Question</u> 5. If M is a homogeneous compact plane continuum, is each proper subcontinuum of M chainable?

REFERENCES

[1] R. Bennett, "Embedding Products of Chainable Continua," Notices of the A. M. S., vol. 11, no. 3 (1964). Abstract 612-9.

[2] R. H. Bing, "A Homogeneous Indecomposable Plane Continuum," Duke Math. J., 15 (1948) pp. 729-742.

[3] ______, "Concerning Hereditarily Indecomposable Continua," Pacific J. Math., 1 (1951) pp. 43-51.

[4] ______, "Snake-like Continua," Duke Math. J., 18 (1951) pp. 653-663.

[5] J. B. Fugate, "On Decomposable Chainable Compact Continua," Notices of the A. M. S., 11, no. 5 (1964). Abstract 614-41.

[6] O. H. Hamilton, "A fixed point Theorem for Pseudo-arcs and Certain other Metric Continua," Proc. A. M. S., 2(1951) pp. 173-174.

[7] A. Lelek, "On Weakly Chainable Continua," Fund. Math., 51 (1962/63) pp. 271-282.

[8] Mioduszowski, "A Functional Conception of Snake-like Continua," Fund. Math.,LI 2 (1962) pp. 179-189.

[9] R. M. Schori, "A Universal Snake-like Continuum," Notices of the A. M. S., 11, no. 3 (1964). Abstract 612-5.

THE EXISTENCE OF A COMPLETE METRIC FOR A SPECIAL MAPPING SPACE AND SOME CONSEQUENCES

by

Louis F. McAuley*

Introduction. Suppose that T is a complete metric space and that K is a compact metric space. The space T^K of all (continuous) mappings f of K into T with the compact-open topology is a complete metric space. There are important applications of special subspaces of T^K. For example, see the work of Dyer and Hamstrom [2]. There they define the concept of a completely regular mapping p from T onto a metric space B. And, they consider the space G_b of all homeomorphisms h_b of K onto $p^{-1}(b)$ for each b in B. It is shown that the space G^* which is the union of the various spaces G_b, $b \in B$, is a complete metric space. This fact is essential in their proof that certain completely regular mappings are locally trivial fiber mappings. The author [3] uses a similar argument to show that a more general class of completely regular mappings have the weak bundle property I with respect to a space K. In this case, the author replaces G_b with the space of all mappings of K onto $p^{-1}(b)$. And, G^* is a complete metric space. When one considers the reverse of this situation, that is, let G_b denote the space of all mappings of $p^{-1}(b)$ into K, then it is no longer apparent that G^* possesses either a metric or a complete metric. We are considering special subspaces of neither T^K nor K^T because the mapping domain varies over the inverse sets $p^{-1}(b)$, $b \in B$. For example, one may wish to consider the case that $p^{-1}(b)$ is homeomorphic to a Cantor set and K to be the interval $[0, 1]$. Here, the space G_b has the useful property of being locally n-connected (in the homotopy sense) for each integer $n \geq 0$. There are other interesting situations where it only makes sense to consider maps from $p^{-1}(b)$ to K —not the other way around. Yet, there seems to be no information available concerning spaces of mappings from subsets of T to K. We show that it is possible to assign a complete metric to certain spaces of mappings from subsets of T to K which is related to the topology of both T and K in a natural way.

---

* Research for this paper was supported in part by NSF-GP-4571.

<u>A Space of Mappings</u>. Suppose that C is a <u>continuous</u> collection of <u>closed</u> subsets of T. That is, each element of C is closed and furthermore, if $\{g_i\}$ is a sequence of elements of C such that $\{g_i\} \to g$, a closed subset of T, then $g \in C$. Notice that the elements of C may intersect contrary to the usual notion of continuous collections.

For each g in C, let G_g denote the space of all mappings of g into K, a complete metric space. Let G denote the collection of all G_g, g in C, and G^* denote the union of elements of G. That is, G^* is the set of all mappings f in G_g for each g in C.

<u>A metric for</u> G^*. If $g \in G^*$, let $\hat{f}$ denote the graph of f in $T \times K$, i.e., the set of all points $(x, f(x))$ in $T \times K$ such that $x \in g$ and f maps g into K. For each pair of elements f, g of G^* where $f \in G_a$ and $g \in G_b$, let $D(f, g) = H(\hat{f}, \hat{g})$ where H denotes the Hausdorff metric on the space of all closed subsets of $T \times K$. That is, let d be a bounded complete metric for $T \times K$. Then $H(\hat{f}, \hat{g}) = \text{lub}\{\varepsilon \mid N_\varepsilon(\hat{f}) \supset \hat{g}$ and $N_\varepsilon(\hat{g}) \supset \hat{f}\}$ where $N_\varepsilon(A)$ denotes the ε-neighborhood of the subset A of T. Clearly, D is a metric for G^* but D may not be complete. To this end, we prove the following theorem.

THEOREM 1. The space G^* is a complete metric space.

Proof. Let $\hat{G}$ denote the collection of all graphs $\hat{g}$ for the various elements g in G. Now, let $\text{Cl}\,\hat{G}$ be the closure of $\hat{G}$ in the space S of all closed subsets of $T \times K$ with the Hausdorff metric H as defined above. As a subspace of S, $\text{Cl}\,\hat{G}$ is a complete metric space. We wish to show that $\hat{G}$ is a G_δ-set (an inner limiting set).

For each element A in $\text{Cl}\,\hat{G}$, there is c in C such that $c \times K$ contains A. This follows from the definition of $\text{Cl}\,\hat{G}$ and the fact that C is a continuous collection. Let A_i denote the collection, if it exists, of all elements A in $\text{Cl}\,\hat{G}$ such that for some x in c (where $c \times K \supset A$), the set $(x, K) \cap A$ has diameter $\geq 1/i$. It follows that for each i, A_i is closed. We wish to show that $\hat{G} = \text{Cl}\,\hat{G} - \bigcup_{i=1}^\infty A_i$. Clearly, if $g \in G^*$, then $\hat{g} \notin A_i$ for any i. To see that $\hat{G} \supset \text{Cl}\,\hat{G} - \bigcup_{i=1}^\infty A_i$, suppose that $\{\hat{g}_i\} \to A$ where $g_i \in G^*$, A is in $\text{Cl}\,\hat{G} - \bigcup_{i=1}^\infty A_i$, $g \in G_{h_i}$ (that is, g_i maps h_i into K, $h_i \in C$). and $\{h_i\} \to h \in C$. If $(x, k) \in A$ and $(y, k) \in A$, then $x = y$. Furthermore, $h \times K \supset A$. For each x in h, there is a sequence $\{x_i\} \to x$ where $x_i \in h_i$ such that $(x_i, g(x_i))$ lies in $N_{1/i}(A)$ and $N_{1/i}(\hat{g}_i) \supset A$. Therefore, there is a sequence $\{(y_i, k_i)\}$ of points in $h \times K$ such that $d[(x_i, g(x_i)), (y_i, k_i)] < 1/i$. Since d is a complete metric for $T \times K$, we may suppose that $\{k_i\} \to k$ without loss of generality. Thus, $\{y_i\} \to x$ and $\{g(x_i)\} \to k$. Let $g(x) = k$. Now, g is a mapping (continuous) of h into K whose graph $\hat{g}$ is A. We have $\hat{G} = \text{Cl}\,\hat{G} - \bigcup_{i=1}^\infty A_i$ and $\hat{G}$ is a G_δ-set. Consequently, G^* possesses a complete metric which leaves its topology invariant.

A Condition Under Which G is Lower Semi-continuous. In applications, it is useful to know that G is lower semi-continuous. Consider the following:

Definition. We shall say that the collection G is regular iff for each g in G and $\varepsilon > 0$, there is a $\delta > 0$ such that if $H(h, g) < \delta$ (where H is a Hausdorff metric for the space of closed subsets of T), then there is a continuous mapping f_{hg} which takes h onto g and moves no point as much as ε. If f_{hg} is a homeomorphism in each case, then G is said to be completely regular.

THEOREM 2. If G is regular, then G is lower semi-continuous. That is, if $\{g_i\} \to g$ where $g_i, g \in G$, then G_g is in the closure of $\bigcup_{i=1}^{\infty} G_{g_i}$.

Proof. Let $f \in G_g$. There is no loss of generality in assuming that there is a mapping $f_{g_i g}$, for each i, which takes g_i onto G and moves no point as much as $1/i$. Now, $t_i = ff_{g_i g}$ maps g_i into K. The graph of $f_{g_i g}$ is the set of points $x, ff_{g_i g}(x))$ whereas the graph of f may be viewed as the set of points $(f_{g_i g}(x), ff_{g_i g}(x))$. Clearly, the sequence of graphs $\{\hat{t}_i\} \to \hat{f}$ and $f \in$ closure of $\bigcup_{i=1}^{\infty} G_{g_i}$.

THEOREM 3. Suppose that for each g in G, G_g is LC^n (in the homotopy sense) and that G is completely regular. Then G is equi-LC^n.

A proof of Theorem 3 is easy to construct if one follows that given for Lemma 1.2 [3] and observes the facts below.

Suppose that $\delta > 0$, $t \in G$, and h is a homeomorphism of s onto t which moves no point as much as δ. Let $N_\delta(f)$ be a δ-neighborhood of f in G_s. For each g in G_s let $\varphi(x, g(x)) = (h(x), g(x))$. That is, φ induces a 1-1 mapping from $G_s \cap N_\delta(f)$ into $G_t \cap N_{2\delta}(f)$. Furthermore, $\varphi^{-1}(y, z(y)) = (h^{-1}(y), z(y))$ for $z \in G_t$ induces a 1-1 mapping from $G_t \cap N_\delta(f)$ into $G_s \cap N_{2\delta}(f)$.

An Application. Suppose that D is the dyadic group. Furthermore, suppose that there is a free action of D on a connected n-manifold M^n. (It is unknown whether a p-adic group for prime p can act freely on an n-manifold.) That is, there is a mapping (continuous) $t\colon D \times M^n \to M^n$ such that (1) $t(e, x) = x$ where e is the identity of D and $x \in M^n$, (2) $t(d_1 d_2, x) = t[d_1, t(d_2, x)]$ for each d_1, d_2 in D and x in M^n (transitive on the left), and (3) if for d in D, $d \neq e$, then $t(d, x) \neq x$ for each x in M^n (free action).

The orbit $OR(x)$ of a point x is the set of all points $t(d, x)$ for each d in D. Furthermore, the orbit space OR is the decomposition space whose points are the orbits $OR(x)$ for the various points x in M^n and whose topology is the usual quotient space topology. The natural mapping $p\colon M^n \to OR$ where $p(y) = OR(x)$ if $y \in OR(x)$, is an open mapping. Also, each orbit $OR(x)$ is homeomorphic to a Cantor set. It also follows easily that p is completely regular.

Now, Yang [4] has shown that the dimension of OR must be either $n+2$ or infinite while Bredon, Raymond, and Williams [1] have shown that it is not infinite. Let $T = M^n$, $B = OR$, and $K = [0, 1]$. Thus, we have a mapping $p\colon T \to B$ which satisfies the hypothesis of Theorem 2 in [3] whose proof rests on the fact that the space G^* of all mappings of $p^{-1}(b)$ into a compact space K for the various b in B is a complete metric space among other things. Consequently, p has the weak bundle property II with respect to $K = [0, 1]$.

Now, let $g_b[p^{-1}(b)]$ be the Cantor set on K for some b in $B = OR$. Now, by Theorem 2[3], there is an open set U_b and a mapping $\varphi_{U_b}\colon p^{-1}(U_b)$ into $U_b \times K$ such that

(a) $\varphi_{U_b} | p^{-1}(b) = g_b$, and

(b) $p | p^{-1}(U_b) = \pi\varphi_{U_b}$ where π is the projection mapping from $U_b \times K$ onto U_b.

We see that the image of an open set $p^{-1}(U_b)$ in M^n may be pulled apart along $\varphi_{U_b}(p^{-1}(b))$ so that there are infinitely many pairwise disjoint open sets in $U_b \times K$ which miss $\varphi_{U_b}(p^{-1}(U_b))$.

Let C be the set of points (u, k) in $U_b \times K$ such that k is the first point of $\varphi_{U_b}(p^{-1}(U_b))$ on $u \times K$, i.e., on $[0, 1]$. Now C is homeomorphic to U_b. Furthermore, $\varphi_{U_b} | \varphi_{U_b}^{-1}(C)$ is a mapping of $\varphi_{U_b}^{-1}(C)$ onto C. If dimension U_b is $n+2$, then $m = \varphi_{U_b} | \varphi_{U_b}^{-1}(C)$ raises dimension by 3 since $\varphi_{U_b}^{-1}(C)$ is at most of dimension $n-1$. Thus, m can not be a countable-to-one open mapping. Although m may not be open, it is similar to an open mapping. Do such mappings raise dimension?

<u>Questions</u>. The space S of all countable-to-one mappings of a Cantor set C into $[0, 1]$ is LC^n for each integer $n \geq 0$. Is S a complete metric space? Does the mapping m described above fail to raise dimension? Affirmative answers to these questions imply that there is no free action of the dyadic group on M^n.

REFERENCES

[1] Bredon, G. E., Raymond, Frank, and R. F. Williams, "p-adic group of transformations," TAMS, vol. 99(1961) pp. 488-498.

[2] Dyer, Eldon, and M. E. Hamstron, "Completely regular mappings," Fund. Math., vol. 45 (1957) pp. 103-118.

[3] McAuley, L. F., "Completely regular mappings, fiber spaces, the weak bundle properties, and the generalized slicing structure properties," This volume.

[4] Yang, C. T., "p-adic transformation groups," Mich. Math. Jour., vol. 7 (1960) pp. 201-218.

RUTGERS UNIVERSITY

FINITE DIMENSIONAL SUBSETS OF INFINITE DIMENSIONAL SPACES

by

David W. Henderson*

This paper gives a summary of some of the results related to the <u>Question</u>: Which spaces of dimension more than n have subsets of dimension n ?

(Throughout this paper, "space" will mean "separable metric space.")

We first state a few appropriate definitions. Let n always denote a non-negative integer. The empty set has dimension -1 (written $\dim \emptyset = -1$). A space X has dimension $\leq n$ ($\dim X \leq n$), if X has a neighborhood basis composed of open sets each with boundary of dimension $\leq n-1$. $\operatorname{Dim} X = n$ (or X is n-dim) if "$\dim X \leq n$" is true and "$\dim X \leq n-1$" is false. $\operatorname{Dim} X = \infty$ (or X is ∞-dim) if "$\dim X = n$" is false for each n. A space X will be called countable-dimensional (c-∞-dim) if $\dim X = \infty$ and X is the union of a countable number of 0-dim spaces. An ∞-dim space will be called <u>strongly infinite dimensional</u> (s-∞-dim) if it is not c-∞-dim.

1. <u>Every space contains compact 0-dim subsets</u>. Each point is such a subset.

2. <u>Each n-dim space contains closed p-dim subsets for all $p \leq n$</u>. This follows directly from the above inductive definition of dimension.

3. (Mazurkiewicz) <u>There exists n-dim spaces, for each n, all of whose compact subsets have dimension zero</u>. These are the n-dim, totally disconnected spaces constructed by Mazurkiewicz [16], because any compact totally disconnected space is 0-dim. (See [11], page 20.)

4. (Hurewicz and Tumarkin) <u>If $2^{\aleph_0} = \aleph_1$, then there is an s-∞-dim space with no c-∞-dim or n-dim ($n > 0$) subsets</u>. Hurewicz showed [10] that there is a space in the Hilbert Cube each finite-dimensional subset of which is countable. (Every countable set is 0-dim [11, page 10].) Tumarkin (the proof is in [38] and an English summary is in [37]) extended this to show that in every s-∞-dim space there is an s-∞-dim space each of whose subsets are either countable or s-∞-dim.

5. (Tumarkin) <u>A compactum contains compact subsets of each finite dimension</u>

---

* The author was partially supported by NSF Contract GP-3857 during preparation of this paper.

if and only if it contains a compact c-∞-dim subset. The proof of this is quite easy and is contained in [38].

6. (Hurewicz and Wallman) If a space is complete and c-∞-dim, then it has closed n-dim subsets, for each n. Hurewicz and Wallman [11, page 50-51] show that every complete, c-∞-dim space has transfinite dimension (defined by extending to all ordinal numbers the inductive definition given in this paper). It is clear that any space with transfinite dimension has closed n-dim subsets, for each n.

7. (Smirnov) If X is c-∞-dim, and there is a map with countable-multiplicity of X into a complete space, then X has closed n-dim subsets, for each n. Smirnov [33] shows that X has transfinite dimension. (See 6.)

8. If the homology dimension of a compact space X is n ($< \infty$) but X is ∞-dim, then X contains no n+1-dim compact subset. (The homology dimension of X is the largest n such that, for some compact set $C \subset X$, $H^n(X,C;Z) \neq 0$ (Cech cohomology).) In [11], pages 151-152, Hurewicz and Wallman show that the homology dimension of a compact n-dim space is n. (This theorem was originally due to Alexandrov [3].) Thus, if a compact space has a compact n-dim subset, then its homology dimension is greater than or equal to n.

9. There is an ∞-dim compact space X with no n-dim compact subsets ($n \geq 1$). A description of this space will appear elsewhere ([8]). It is known that some component of a compact n-dim space must be n-dim (see [11, page 94]). As was noted in 2 above, every compact n-dim space contains a compact 1-dim subset. Therefore, in order to show that the space X has no n-dim compact subsets ($n \geq 1$) it is only necessary to show that X contains no 1-dim continua. (A continuum is a compact connected space.) In [8] it is shown that for every non-degenerate continuum K in X, there is a compact set F and a map f, such that the hypotheses of the following lemma are satisfied.

LEMMA. If K is a compact subset of the compact space F and $f\colon F \to I^2$ ($I^2 = [0, 1] \times [0, 1]$) is a map such that

(*) if G is a compact subset of F that separates $f^{-1}(I \times 1)$ from $f^{-1}(I \times 0)$, then $G \cdot X$ contains a continuum joining $G \cdot f^{-1}(0 \times I)$ to $G \cdot f^{-1}(1 \times I)$,

then $\dim(X) \geq 2$.

Proof. Suppose that $\dim(K) < 2$. Then there is a map $g\colon K \to \operatorname{Bd} I^2$, such that $g|K \cdot f^{-1}(\operatorname{Bd} I^2) = f|K \cdot f^{-1}(\operatorname{Bd} I^2)$. (See [11, page 83].) We may extend g to $G\colon U \to \operatorname{Bd} I^2$, where U is a neighborhood of $K + f^{-1}(\operatorname{Bd} I^2)$ in F and $G|f^{-1}(\operatorname{Bd} I^2) = f|f^{-1}(\operatorname{Bd} I^2)$. Let

$$A = (I \times 0) + (1 \times [0,\tfrac{1}{2})) + (0 \times [0,\tfrac{1}{2})),$$

$$B = (0 \times (\tfrac{1}{3}, 1]) + ([0, 1) \times 1),$$

$$C = (1 \times (\tfrac{1}{3}, 1]) + ((0, 1] \times 1).$$ (See figure.)

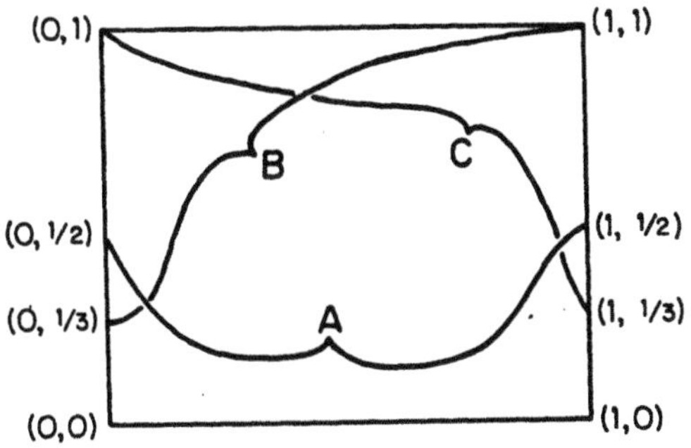

$G^{-1}(A)$ is a neighborhood of $f^{-1}(I \times 0)$ and thus $N = $ (Boundary of $G^{-1}(A)$ in F) is a compact set separating $f^{-1}(I \times 0)$ from $f^{-1}(I \times 1)$. It can be seen that $N \cdot f^{-1}(0 \times I) \subset f^{-1}(0, \frac{1}{2}) \subset G^{-1}(B)$ and $N \cdot f^{-1}(1 \times I) \subset f^{-1}(1, \frac{1}{2}) \subset G^{-1}(C)$. Also $G^{-1}(A) \cdot G^{-1}(B) \cdot G^{-1}(C) = \emptyset$ since $A \cdot B \cdot C = \emptyset$. Therefore, $R = N - (G^{-1}(B) + G^{-1}(C))$ is a closed (possibly empty) subset of N that separates $N \cdot f^{-1}(0 \times I)$ from $N \cdot f^{-1}(1 \times I)$. But this contradicts (*), since $K \subset G^{-1}(A) \cdot G^{-1}(B) \cdot G^{-1}(C)$. This completes the proof of the Lemma.

10. Wrong Results. In this section we will mention two results which have been claimed in print, but whose proofs are faulty.

Van Heemert claimed ([40]) in 1946 to have shown that each compact space contains 1-and 2-dim compact subsets. The example mentioned in 9 shows that his result is incorrect. The main error is Van Heemert's use ([40, pages 568-9]) of a theorem ([7], Seite 229, Satz 4) of Freudenthal's. Freudenthal's theorem deals with inverse limit (R_n-adic) sequences which have the property that all finite compositions of the bonding maps are normal and irreducible (see [7] or [40] for definitions). Van Heemert assumes only that the bonding maps (between consecutive polyhedra in the inverse limit sequence) are normal and irreducible. It can be shown that (a) Van Heemert's modification of Freudenthal's theorem is false and (b) Van Heemert's arguments will not work if the correct version of the Freudenthal theorem is used.

The author recently published an abstract ([9]) in which he claimed to have proven that every compact non-zero-dimensional space contains a connected 1-dim subset. However, his proof was incorrect.

11. We finish the paper with a conjecture.

Conjecture: Every compact space contains connected subsets of every dimension less than the dimension of the space.

INSTITUTE FOR ADVANCED STUDY

REFERENCES

(These references include most of the published works pertaining to the dimension theory of infinite dimension spaces. The square brackets at the end of some of the references refers to the review of the article in Mathematical Reviews (MR). Russian journals are given by the abbreviations used in MR and also by the standard Russian abbreviations.)

[1] Alexandrov, P. S., "The present status of the theory of dimension," Amer. Math. Soc. Translations, Series 2, vol. 1 (1955) pp. 1-26.

[2] _______, "On some results concerning topological spores and their continuous mappings," General Topology and its Relations to Modern Algebra and Analysis (Proc. Sympos., Prague, 1961), pp. 41-54. Academic Press, New York; [MR 26(3003)].

[3] _______, "Dimensionstheorie," Math. Ann. 106 (1932) pp. 161-238.

[4] Anderson, R. D., Connell, E. H., "Concerning closures of images of the reals," Bull. Acad. Polon. Sic. Sér. Sic, Math, Astrom, Phys., 9 (1961) pp. 807-10, [MR 24(A2369)].

[5] Arhangle'skii, A., "The ranks of systems of sets and the dimensions of spaces," Soviet Math. Dokl. 3(1962) pp. 456-459 [MR 24(A2368)]. (The details in Fund. Math. 52(1963) pp. 257-275.)

[6] Erdős, Paul, "The dimension of rational points in Hilbert Space," Ann. Math. 41 (1940) pp. 734

[7] Freudenthal, H., "Entwicklungen von Raumen and ihren Gruppen," Compositio Mathematica 4 (1937) pp. 10-234.

[8] Henderson, D. W., "An infinite dimensional compactum with no positive dimensional compact subsets, (submitted).

[9] _______, "Every compactum contains a connected 1-dimensional subset," Abstract 622-69, Notices of AMS, 12 (1965) p. 347. (The proof claimed in this abstract is faulty.)

[10] Hurewicz, W., "Une remarque sur l'hypotheses du continie," Fund. Math. 19 (1932) pp. 8-9.

[11] Hurewicz and Wallman, <u>Dimension Theory</u>, Princeton University Press, Princeton, N. J., 1941.

[12] Levsenko, B. T., "Strongly infinite-dimension spaces " (Russian), Vestnik Moskov Univ. Ser. Mat. Astr. Fiz Nim., (Becthmk M.Y.C.M.A.Φ.H.) (1959), no. 5, 219 [MR 22(4053)].

[13] _______, "On ∞-dimensional spaces," Soviet Math. Dokl., 2 (1961) p. 915 [MR 24(A1708)].

[14] Mardesic, Sibe, "Covering dimension and inverse limits of compact spaces," Illinois J. of Math., 4 (1960) pp. 278-291.

[15] Morita, K., "On closed mappings and dimension," Proc. Japan Acad., 32 (1956) pp. 161-165.

[16] Mazurkiewicz, "Sur les problèmes κ et λ de Urysohn," Fund. Math., 10 (1927) pp. 311-319.

[17] Magata, J., "On the countable dimensional spaces," Proc. Japan Acad. 34 (1958) pp. 146-149.

[18] ______, "On the countable sum of 0-dimensional metric spaces," Fund. Math., 48 (1960) pp. 1-14 [MR 22(5028)].

[19] ______, "Two remarks on dimension theory for metric spaces," Proc. Japan Acad., 36 (1960), 53 [MR 22(9963)].

[20] Nagama, Ketô, "Finite-to-one closed mappings and dimension I," Proc. Japan Acad., 34 (1958) pp. 503-506 [MR 21(862)].

[21] ______, "Finite-to-one closed mappings and dimension II," Proc. Japan Acad., 35(1959) pp. 437-439 [MR 22(4025)(1961)].

[22] ______, "Mappings of finite order and dimension theory," Japan J. Math., 30(1960) pp. 25-54 [MR 25(5494)]. (This paper contains a general development of dimension theory for metric spaces.)

[23] Pasynkov, B., "Absence of polyhedral spectra for bicompacta," Soviet Math. Dokl., 3(1962) pp. 113-116.

[24] ______, "On the spectra and dimensionality of topological spaces," Ma. Sv. (N.S.)(Ma. C. H. C.) 57 (99)(1962), pp. 449-476 [MR 26(1856)]. (This paper contains details and proofs of the preceeding paper and more.)

[25] Šersnev, M., "Strong dimension of mappings and the associative characterization of dimension for arbitrary metric spaces," Soviet Math. Dokl., 1(1961) pp. 1267-1269 [MR 23(A2865)]. (This details in Ma. Sb. (N.S.)(Ma. C. H. C)) 60 (102) (1963) pp. 207-218 [MR 28(585)].

[26] Sklyarenko, E. G., "Two theorems on infinite-dimensional spaces," Soviet Math. Dokl., 3(1962) pp. 547-550 [MR24(A2367)].

[27] ______, "On dimensional properties of infinite dimensional spaces," Izv. Akad. Nauk SSSR ser Mat (NAHCCCP) 23 (1959) p. 197 MR 21[5179] (translation: AMS Trans. (2) 21 (1962) p. 35-50).

[28] ______, "Some remarks on spaces having an infinite number of dimensions" (Russian), Dokl. Akad. Nauk SSSR (DAHCCCP) 126 (1959) p 1203 MR 22[1885].

[29] ______, "Homogeneous spaces of an infinite number of dimensions," Soviet Math. Dokl., 2(1961) p. 1569.

[30] Smirnov, Ju. M., "Some remarks on transfinite dimension," Dokl. Akad. Nauk SSSR (DAHCCCP) 141(1961) p. 814 [MR 26(6934)]. (Translated in Soviet Math. Dokl., 2(1961) p. 1572.)

[31] ______, "On dimensional properties of infinite dimensional spaces," Gen. Top. etc., (Proc. Symposium Prague, 1961) pp. 334-336.

[32] ______, "Universal spaces for certain classes of ∞-dim spaces," Izd. Akad. Nauk SSSR, Sem. Mat. (NAHCCP) 23(1959) p. 185 [MR 5178]. (Translated in AMS translations (2) 21 (1962) pp. 21-33.)

[33] ______, "On transfinite dimension" (Russian), Mat. Sb. (W.S.) (Mat. C. H. C.) 58 (100)(1962) pp. 415-522.

[34] Smirnov and Sklyarenko, "Some questions in dimension theory" (Russian) Proc. 4th All-Union Math. Congress (Leningrad 1961), vol. I, pp. 219-226. (T

[35] Toulmin, G. H., "Shuffling ordinals and transfinite dimension," Proc. London Math. Soc., 4 (1954) pp. 177-195.

[36] Tumarkin, L. A., "On the decomposition of spaces into a countable number of 0-dimensional sets" (Russian), Vestnik Moskow, Univ. Sec. I Mat. Mek, (Becthnk M. Y., C. I) 1960, no. 1, pp. 25-32 [MR 25(554)].

[37] ______, "Concerning infinite-dimensional spaces," General Topology and its Relations to Modern Algebra and Analysis (Proc. Sympos., Prague, 1961) pp. 352-353.

[38] ______, "On strongly and weakly infinite-dimensional spaces" (Russian), Vestnik. Moslow Univ. Ser I, Mat. Meh. (Becthnk M. Y. C. I.)(1963) no. 5, pp. 24-27 [MR 27(5230)].

[39] ______, "On infinite-dimensional Cantor manifolds" (Russian), Dokl. Akad. Nauk SSSR (DAHCCCP) 115(1957) pp. 244-246.

[40] Van Heemert, A., "The existence of 1-and 2-dimensional subspaces of a compact metric space, Amsterdam Acad. van Wetenschappen-Indagations Mathematicae, 8 (1946) pp. 564-569. (This paper is incorrect.)

[41] Alexandroff, P. S., "Results in topology in the last 25 years," Russian Math. Surveys 15 (1960).

[42] ______, "On some basic directions in general topology," Russian Math. Surveys 19 (1964) # 6, pp. 1-39.

[43] Sklyorenko, E. G., "Representations of infinite-dimension compacta as an inverse limit of polyhedra," oviet Math. Dokl. 1 (1960) pp. 1147-49.

[44] Nogata, J., Modern Dimension Theory, Interscience, New York, 1965.

TYPES OF ULTRAFILTERS

by

Mary Ellen Rudin*

Introduction.

If S is a set and Ω is a collection of non-empty subsets of S, then Ω is said to be an ultrafilter on S provided that:

(a) the intersection of any two members of Ω belongs to Ω,

and (b) Ω is maximal with respect to (a).

There is a close connection between ultrafilters on N (the set of all positive integers) and the Cech compactification βN of N as the following construction of βN (see [1]) shows:

(i) Let the ultrafilters on N be the points of βN.

(ii) Identity each $n \in N$ with the ultrafilter on N which consists of all subsets of N which contain n.

(iii) For every set $E \subset N$, declare the set of all $\Omega \in \beta N$ such that $E \in \Omega$ to be open in βN; and every open set in βN is a union of such sets.

My concern will be with the space $N^* = \beta N - N$ and a classification of the points of N^*. The points of N^* are called free ultrafilters on N; two free ultrafilters on N are said to be of the same type if there is a homeomorphism of N^* onto N^* which carries one of these ultrafilters onto the other. While the homeomorphisms of βN are easily described (they are induced in a natural way by the permutations of N; see [1]), the situation is more complicated in N^*.

The hypothesis of the continuum will be assumed throughout this paper and c will denote the power of the continuum.

A point p of N^* is called a P-point of N^* if the intersection of every countable collection of neighborhoods of p (in N^*) contains a neighborhood of p. It is shown in [1] that N^* has 2^c P-points if the hypothesis of the continuum is true, and that all P-points are of the same type.

This paper will classify the limit points of countable sets of P-points. The classification is in terms of ultrafilters on the P-points

---

* Work on this paper was supported by the National Science Foundation under grant GP-3857.

themselves. Also it will imply that there are 2^c distinct types of ultrafilters among the limit points of each infinite countable set of P-points. (The results of [1] showed only that there are at least two types of free ultrafilters.)

However, the problem of classifying all of the points of N^* is still far from complete. For instance, I know that there are non-P-points which are not limit points of any countable set of P-points. And I do not know whether there are any points of N^* except P-points which are not limit points of any countable set. Also I do not know if there are any points of N^* which are not limit points of any countable discrete set but are limit points of some countable set.

Classification of 2^c Types of Points in N^*.

1. Definition. For every $E \subset N$, put $W(E) = \{\Omega \in N^*: E \in \Omega\}$. The sets $W(E)$ are the open-closed sets of N^* and form a basis for the open sets in N^*. Observe that $W(E_1) \subset W(E_2)$ if and only if $E_1 - E_2$ is finite.

2. LEMMA. If $X \cup Y$ is a countable discrete subset of N^* and Ω is in the closures of both X and Y, then Ω is in the closure of $X \cap Y$.

Proof. Suppose on the contrary that there is an $E \in \Omega$ such that $W(E) \cap (X \cap Y) = \emptyset$. Since no point of $X \cup Y$ is a limit point of $X \cup Y$ and $X \cup Y$ is countable, there are disjoint sets $E_\theta \in \theta$ for each $\theta \in X \cup Y$. Then define $E_X = \{n \in N: n \in (E_\theta \cap E)$ for some $\theta \in (X \cap W(E))\}$ and define $E_Y = \{n \in N: n \in (E_\theta \cap E)$ for some $\theta \in (Y \cap W(E))\}$. Since Ω is a limit point of X and $\Omega \in W(E)$, Ω is a limit point of $X \cap W(E)$; and since $W(E_X)$ is an open closed set containing $X \cap W(E)$, $\Omega \in W(E_X)$. Similarly $\Omega \in W(E_Y)$. But since $E_X \cap E_Y = \emptyset$, $W(E_X) \cap W(E_Y) = \emptyset$ which is a contradiction.

3. Definition. If X is a countable discrete subset of N^* and $p \in \overline{X}$, then define $\Omega_X(p) = \{Y \subset X: p \in \overline{Y}\}$. By Lemma 2, $\Omega_X(p)$ is an ultrafilter on X whenever p is a limit point of X.

4. LEMMA. If X is a countable, infinite, discrete subset of N^*, then $\overline{X} = \beta X$, $\overline{X}$ is homeomorphic to βN, and $\overline{X} - X$ is homeomorphic to N^*.

Proof. Let π be any one-to-one correspondence from X onto N and, for $p \in \overline{X}$ define $f(p) = \{E \subset N: \pi(E) \in \Omega_X(p)\}$. Clearly f is a homeomorphism of $\overline{X}$ onto βN and $f(X) = N$.

5. Definition. If Ω_1 and Ω_2 are ultrafilters on countable sets X and Y, respectively, then Ω_1 and Ω_2 are said to be similar if there is a one-to-one correspondence f of X onto Y such that, for each $E \subset X$, $E \in \Omega_1$ if and only if $f(E) \in \Omega_2$. The main theorem can now be stated.

6. THEOREM. Suppose that X and Y are countable sets of P-points and that p and q are limit points of X and Y, respectively. Then p and q are of the same type if and only if $\Omega_X(p)$ and $\Omega_Y(q)$ are similar.

Observe that every countable set of P-points is discrete.

Then let us immediately point out one consequence of this theorem. Observe that X has 2^c limit points (since N^* has 2^c points this follows from Lemma 4). Also, since there are only c permutations of X, each ultrafilter on X is similar to at most c other ultrafilters on X. Therefore there are 2^c distinct types of limit points of X.

7. Proof of the necessity for Theorem 6. Suppose that h is a homeomorphism of N^* onto N^* such that $h(p) = q$.

Select subsets X_1 and Y_1 of X and Y, respectively, such that $X - X_1$ and $Y - Y_1$ are infinite and $p \in \overline{X}_1$ and $q \in \overline{Y}_1$. Let $Y_2 = h(X_1) \cap Y_1$ and $X_2 = h^{-1}(Y_2)$. By Lemma 2, $q \in \overline{Y}_2$ and hence $p \in \overline{X}_2$. Clearly h induces a one-to-one correspondence between $\{Z \subset X_2: p \in \overline{Z}\}$ and $\{V \subset Y_2: q \in \overline{V}\}$. Let g be any one-to-one correspondence from $X - X_2$ onto $Y - Y_2$. Now define f as the one-to-one correspondence from X onto Y such that $f(x) = h(x)$ for $x \in X_2$ and $f(x) = g(x)$ for $x \in X - X_2$. If $E \subset X$, then $p \in \overline{E}$ if and only if p is a limit point of $E \cap X_2$; hence $p \in \overline{E}$ if and only if q is a limit point of $E \cap Y_2$. Therefore, $\Omega_X(p)$ is similar to $\Omega_Y(q)$ since if $E \subset X$, $p \in \overline{E}$ if and only if $y \in \overline{f(E)}$

8. I think the following theorem is interesting in itself and it will be used in the proof of the sufficiency for Theorem 6. The proof of Theorem 8 is similar to the proof of Theorem 4.7 of [1] and especially makes use of Lemma 11 which is proved in [1] although stated in a different form.

Throughout the rest of the paper $\mathfrak{O}$ is used to denote the set of all open-closed subsets of N^*.

A subcollection of $\mathfrak{O}$ will be called a ring if it is closed under finite unions, finite intersections and complements.

9. THEOREM. Assume that for each $n \in N$, h_n is a homeomorphism of N^* onto N^* and that $\{U_n\}$ and $\{h_n(U_n)\}$ are countable families of disjoint open-closed subsets of N^*. Then there is a homeomorphism h of N^* onto N^* such that, for each $n \in N$ and $x \in U_n$, $h(x) = h_n(x)$.

Proof. Define R_1 as the ring generated by $\{U_n\}$. That is A belongs to R_1 if and only if A is $\emptyset$ or N^* or the finite union of U_n's or the finite intersection of $(N^* - U_n)$'s. Let f_1 be the function from R_1 into $\mathfrak{O}$ which preserves unions, intersections, and complements such that $f_1(\emptyset) = \emptyset$ and $f_1(U_n) = h_n(U_n)$.

Using the hypothesis of the continuum, index $W = \{W_\alpha\}$ by the countable ordinals. Lemma 10 which follows will show that f_1 can be extended by transfinite induction to a permutation f_c on $\mathfrak{W}$ such that (1) f_c preserves intersections, unions, and complements and (2) if $V \subset U_n$, $f_c(V) = h_n(V)$. Hence f_c induces a homeomorphism h of N^* onto N^* such that $h(U_n) = h_n(U_n)$.

10. LEMMA. Given $W = W_\alpha \in \mathfrak{W}$, assume that R is a countable ring containing R_1 and that $f\colon R \to \mathfrak{W}$ is an extension of f_1 such that

(1) f preserves intersections, unions, and complements

and (2) if $V \subset U_n$ for any n and $V \in R$, then $f(V) = h_n(V)$.

Then there is a countable ring R' containing $R \cup \{W\}$ and an extension f' of f to R' such that (1) and (2) are satisfied when f and R are replaced by f' and R' and W also belongs to the range of f'.

Proof. First define $Q_n = W \cap U_n$ for each positive integer n; and define A as the ring generated by $\{Q_n\} \cup R$. Clearly A is countable. Let $g\colon A \to \mathfrak{W}$ denote the function which preserves unions, intersections, and complements such that $g(V) = f(V)$ for all $V \in R$ and $g(Q_n) = h_n(Q_n)$ for all $n \in N$.

Since $\emptyset$ and N^* belong to R_1 and hence to A, by Lemma 11 which follows g can be extended to a function $g'\colon (A \cup \{W\}) \to \mathfrak{W}$ which preserves inclusion and proper inclusion. Now define R' to be the ring generated by $A \cup \{W\}$ and f' to be the extension of g' to R' which preserves unions, intersections and complements. Then R' and f' satisfy (1) and, since f' is an extension of g, they also satisfy (2).

Similarly W can be added to the range of f'.

11. LEMMA. Suppose that A is a countable subset of $\mathfrak{W}$, that $\emptyset$ and N^* belong to A and that $g\colon A \to \mathfrak{W}$ preserves inclusion and proper inclusion. Then, if $W \in \mathfrak{W}$, g can be extended to a function $g\colon (A \cup \{W\}) \to \mathfrak{W}$ which also preserves inclusion and proper inclusion.

12. Proof of the sufficiency for Theorem 6. Assume that h is a one-to-one correspondence between X and Y such that for each $E \subset X$, p is a limit point of E if and only if q is a limit point of $h(E)$.

Let $X = \{x_1, x_2, \ldots\}$. Since x_n and $h(x_n)$ are of the same type there is a homeomorphism h_n of N^* onto N^* such that $h_n(x_n) = h(x_n)$. Since X and Y are discrete there are families $\{H_n\}$ and $\{K_n\}$ of disjoint open-closed subsets of N^* where $x_n \in H_n$ and $h(x_n) \in K_n$ for all $n \in N$. Now define $U_n = H_n \cap h_n^{-1}(K_n)$. Observe that $x_n \in U_n$ and that $\{U_n\}$ are disjoint open-closed subsets of N^* as are $\{h_n(U_n)\}$.

Apply Theorem 9 and find a homeomorphism h' of N^* onto N^* such that $h'(U_n) = h_n(U_n)$ and hence $h'(x_n) = h_n(x_n) = h(x_n)$. Observe that if U is any open-closed set in N^* containing p, then p is a limit point of $X \cap U$ and q is a limit point of $h'(X \cap U)$, hence $q \in h'(U)$, hence $h'(p) = q$. But this means that p and q are of the same type. Therefore the proof of our theorem is complete.

13. In the proof of the sufficiency for Theorem 6 we used the fact that X and Y were discrete and that, for $x \in X$, x and $h(x)$ were of the same type; but no other use was made of the fact that X and Y were made up of P-points. Hence the following.

COROLLARY. Suppose X is a countable discrete subset of N^*, p is a limit point of X in N^*, and $q \in N^*$. Then p and q are of the same type if and only if there is a countable discrete subset Y of N^* such that q is a limit point of Y and there is a one-to-one correspondence h from X onto Y which induces a similarity between $\Omega_X(p)$ and $\Omega_Y(p)$ where for $x \in X$, x and $h(x)$ are of the same type.

An example of a non-P-point which is not a limit point of any countable set of P-points.

For $E \subset N$, let $\varphi(E, n)$ be the number of integers between 1 and n which belong to E; then define the density $D(E)$ to be $\lim_{n\to\infty} (\varphi(E, n))/n$ if the limit exists.

Let I denote the set of all subsets of N having density 1. Since I is closed under finite intersection, the set M of all ultrafilters on N having I as a subset is nonvacuous. Observe that a point of M has no element of density 0 and that a point of $N^* - M$ has an element of density 0.

First we show that no point of M is a P-point; suppose that Ω is an ultrafilter on N. Observe that, for $n \in N$, there is one (and only one) $m \leq n$ such that the set of all positive integers which are m modulo n belongs to Ω. Hence, for $n \in N$, there is an element E_n of Ω of density $1/n$. Since $\Omega \in W(E_n)$ for each n, if Ω is a P-point there is a set $E \in \Omega$ such that $W(E) \subset W(E_n)$ for all n. Therefore, since E has density 0, if Ω is a P-point, $\Omega \notin M$.

We now show that no point of M is the limit point of a countable number of P-points by showing that no point of M is a limit point of any countable subset of $N^* - M$. So suppose that $\{\Omega_n\}$ is a countable subset of $N^* - M$ and, for each n select an element E_n of Ω such that $D(E_n) = 0$. We can modify the E_n by deleting a finite number of terms from each so that the union E of the resulting sets has density 0. Then $W(E)$ is an open-closed set in N^* which contains Ω_n for all $n \in N$. But, since $D(E) = 0$, $W(E) \cap M = 0$; thus no point of M is a limit point of $\{\Omega_n\}$.

REFERENCE

[1] Rudin, W., "Homogeneity problems in the theory of Čech compactifications Duke Math. J., vol. 23 (1955) pp. 409-420.

Univ. of Wisconsin

ADDITIONAL QUESTIONS
ON ABSTRACT SPACES

1. (F. B. Jones) Since certain logicians seem to feel that the assumption of the existence of a Souslin space can lead to no difficulties, perhpas the following questions take on additional interest.

If a Souslin space exists, then there is no homeomorphism from it to the real numbers. What other kind of real function exists, or fails to exist? What kind of continuous functions from a Souslin space to itself exist? In particular, is a Souslin space homogeneous?

2. (A. C. Connor) Can the space of all subcontinua of the pseudo-arc be embedded in E^3?

CHAPTER IV: N-MANIFOLDS

TAMING POLYHEDRA IN THE TRIVIAL RANGE*

John L. Bryant**

1. Introduction. One of the major imbedding problems in topology is to determine the equivalence classes of the set of imbeddings of one topological space into another under some suitable definition of equivalence of two imbeddings. In the special case in which the domain space is a polyhedron of dimension k and the imbedding space is a combinatorial n-manifold (without boundary), it has been shown [1, 10] that, whenever $2k + 2 \leq n$, any two "sufficiently close" piecewise linear imbeddings are equivalent by an ambient isotopy. Gluck [7, 8], Greathouse [9], and Homma [11] showed that, under the same conditions, any locally tame imbedding is equivalent to a piecewise linear imbedding by a homeomorphism of the manifold onto itself. Indeed, Gluck [7,8] went further to show that this equivalence can actually be realized by an ambient isotopy.

The purpose here is to show that, for $2k + 2 \leq n$, the set of locally tame imbeddings of a k-dimensional polyhedron into a combinatorial n-manifold is actually larger than it appears to be at first. The main theorem is

THEOREM 1. Suppose f is an imbedding of a k-dimensional polyhedron X^k into a combinatorial n-manifold M^n, where $2k + 2 \leq n$, and P is a tame polyhedron in M^n, with $\dim P \leq \frac{n}{2} - 1$, such that $f|X^k - f^{-1}(P)$ is locally tame. Then f is ε-tame.

2. Definitions. A k-dimensional polyhedron is the underlying space of some finite k-dimensional simplicial complex and will usually be denoted by X^k. A combinatorial n-manifold is a locally finite simplicial complex for which the link of each vertex is combinatorially equivalent to the boundary of the standard n-simplex (that is, only manifolds without boundary are to be considered) and will be denoted by M^n.

An imbedding $f\colon X^k \to M^n$ is said to be tame if there is a homeomorphism h of M^n onto itself such that $hf\colon X^k \to M^n$ is piecewise linear. An imbedding $f\colon X^k \to M^n$ is locally tame at a point x if there is a

---

* The author is indebted to Professor C. H. Edwards, Jr., for his suggestions and assistance.

** NSF Cooperative Graduate Fellow.

polyhedral neighborhood of x on which f is tame. An imbedding will be called ε-<u>tame</u> if for each $\varepsilon > 0$ there is an ε-push h of $(M^n, f(X^k))$ (see §3) such that $hf: X^k \to M^n$ is piecewise linear.

3. <u>Dense and solvable sets of imbeddings</u>. The techniques employed to prove Theorem 1 are based upon the following notions introduced by Gluck in [7].

If A is a closed subset of the topological manifold M, then an ε-<u>push</u> h <u>of the pair</u> (M, A) is an ε-homeomorphism of M onto itself satisfying

1) $h|M - U_\varepsilon(A)$ is the identity, and
2) h is ε-isotopic to the identity by an isotopy $h_t (t \in I)$ such that $h_t|M - U_\varepsilon(A)$ is the identity for each $t \in I$.

Now let M be a topological manifold with complete metric d. Suppose X is a metric space and A is a subset of X such that $\bar{A}$ is compact. $Hom(X, A; M)$ will denote an arbitrary set of imbeddings of X into M all of which agree on $X - A$. $Hom(X, A; M)$ then becomes a metric space by defining

$$d(f, g) = \sup_{x \in A} d(f(x), g(x))$$

for $f, g \in Hom(X, A; M)$.

A subset F of $Hom(x, A; M)$ is said to be <u>dense</u> in $Hom(X, A; M)$ if for each $\varepsilon > 0$ and each $g \in Hom(X, A; M)$, there is an element $f \in F$ with $d(f, g) < \varepsilon$. A subset F of $Hom(X, A; M)$ is said to be <u>solvable</u> if for each $\varepsilon > 0$, there is a $\delta = \delta_F(\varepsilon) > 0$ such that whenever $f, f' \in F$ and $d(f, f') < \delta$, there is an ε-push h of $(M, f(A))$ such that $hf = f'$.

By using Homma's techniques [11], Gluck [7, 8] was able to prove

THEOREM 2. The union of two dense, solvable subsets of $Hom(X, A; M)$ is dense and solvable.

4. <u>Lemmas</u>. By applying Gluck's results [7, 8], one can easily prove

LEMMA 3. Suppose f is an imbedding of X^k into Euclidean n-space E^n $(2k + 2 \leq n)$ and P is a polyhedron piecewise linearly imbedded in E^n, with $\dim P \leq \frac{n}{2} - 1$, such that $f|X^k - f^{-1}(P)$ is locally tame. Then for each $\varepsilon > 0$, there is an ε-push h of $(E^n, f(X^k - f^{-1}(P)))$ such that $hf|X^k - f^{-1}(P)$ is locally piecewise linear and $h|P$ is the identity.

Suppose f is an imbedding of X^k into E^n as described in the hypothesis of Lemma 3. By applying Lemma 3, one may assume that $f|X^k - Y$ is locally piecewise linear, where Y denotes $f^{-1}(P)$. Let $Hom(X^k, X^k - Y; E^n)$ be the set of all imbeddings of X^k into E^n which agree with f on Y.

Consider the set F of all f' in $\mathrm{Hom}(X^k, X^k - Y; E^n)$ for which there is a polyhedron Z in $X^k - Y$ such that

1) $f'|X^k - Z = f|X^k - Z$, and
2) $f'|Z$ is piecewise linear.

LEMMA 4. The set F is a dense, solvable subset of $\mathrm{Hom}(X^k, X^k-Y; E^n)$.

Proof. If $\varepsilon > 0$ and $g \in \mathrm{Hom}(X^k, X^k - Y; E^n)$ are given, choose a polyhedron Z in $X^k - Y$ such that

$$d(f(x), g(x)) < \varepsilon$$

for all $x \in X^k - Z$. By using standard extension theorems and general position arguments, one can extend $f|X^k - Z$ to an imbedding f' of X^k into E^n so that $f'|Z$ is piecewise linear and $d(f', g) < \varepsilon$. By definition, $F' \in F$; hence, F is dense.

To see that F is solvable, choose $\delta_F(\varepsilon) = \varepsilon$ for any arbitrary positive number ε, and suppose that f_0, $f_1 \in F$ with $d(f_0, f_1) < \varepsilon$. Notice that f_0 and f_1 agree with f except on a polyhedron Z in $X^k - Y$, and that both f_0 and f_1 are peicewise linear on Z. By a suitable modification of a theorem of Bing and Kister [1], one obtains an ε-push h of $(E^n, f_0(Z))$ such that

1) $hf_0|Z = f_1|Z$ and
2) $h|P \cup f(X^k - Z) = 1$.

Thus F is solvable.

5. A special case.

THEOREM 5. Suppose Y is a subpolyhedron of X^k and f is an imbedding of X^k into E^n $(2k + 2 \leq n)$ such that $f|Y$ is locally tame and $f|X^k - Y$ is locally tame. Then f is ε-tame.

Proof. Since Gluck's results [7, 8] and Lemma 3 allow one to assume that $f|Y$ is piecewise linear and $f|X^k - Y$ is locally piecewise linear, the theorem follows immediately from Lemma 4 and Theorem 2, for the set of piecewise linear extensions of $f|Y$ in $\mathrm{Hom}(X^k, X^k-Y; E^n)$ is known to be dense and solvable.

Using Theorem 5, one can prove, by induction on k, the following theorem, which has been established for $k = 1$ and $k = 2$ by Cantrell [3] and Edwards [6], respectively.

THEOREM 6. A k-complex K in E^n $(2k + 2 \leq n)$ is ε-tame if, and only if, each simplex of K is tame.

From this and a theorem of Klee [12], one obtains an alternative proof of a theorem of Bing and Kister [1].

THEOREM 7. If f is an imbedding of X^k into the n-plane E^n in E^{n+k} $(n \geq k + 2)$, then $f\colon X^k \to E^{n+k}$ is ε-tame.

6. Proof of Theorem 1.

Case 1. $M^n = E^n$. Let $Y = f^{-1}(P)$. One may assume that P is piecewise linearly imbedded in E^n, by Gluck's results [7, 8], and that $f|X^k - Y$ is locally piecewise linear from Lemma 3.

Let Φ denote the set of all piecewise linear imbeddings of P into E^n. For each $\varphi \in \Phi$, choose an imbedding f_φ of X^k into E^n such that

1) $f_\varphi|Y = \varphi f|Y$,
2) $f_\varphi|X^k - Y$ is locally piecewise linear, and
3) $fi = f$ where i is the inclusion of P into E^n,

and let $\mathrm{Hom}_\varphi(X^k, X^k - Y; E^n)$ be the set of all imbeddings of X^k into E^n which agree with f_φ on Y. If, for each $\varphi \in \Phi$, one defines F_φ to be the subset of $\mathrm{Hom}_\varphi(X^k, X^k - Y; E^n)$ associated with f_φ as described in §4, it follows that $F = \cup_{\varphi \in \Phi} F_\varphi$ is a dense, solvable subset of $\mathrm{Hom}(X^k; E^n)$, the set of all imbeddings of X^k into E^n. Thus, Theorem 1 is proved in the case $M^n = E^n$, since the set of all piecewise linear imbeddings of X^k into E^n is dense and solvable in $\mathrm{Hom}(X^k; E^n)$.

Case 2. (General Case). Suppose $x \in X^k$. Let U be an open combinatorial n-ball in M^n containing $f(x)$ and let Z be a polyhedral neighborhood of x in X^k such that $f(Z) \subset U$. Choose a piecewise linear homeomorphism g of U onto E^n. Then, by case 1,

$$gf|Z\colon\ Z \to E^n$$

is tame so that

$$f|Z\colon\ Z \to M^n$$

is tame, also. Hence, f is a locally tame imbedding of X^k into M^n so that, by Gluck's results [7, 8], f is ε-tame.

THEOREM 8. Theorems 5 and 6 remain valid if E^n is replaced by an arbitrary M^n.

7. Further Applications. If M^n is a piecewise linear submanifold of M^{n+k}, then M^n is also locally flat in M^{n+k} whenever $k = 1$ (Brown [2]) or $k \geq 3$ (Zeeman [13]).

Thus, Theorem 7 and a theorem of Klee [12] can be used to prove

THEOREM 9. If M^n is piecewise linearly imbedded in M^{n+k} ($n \geq 2k+2$) and if M^n is assumed to be locally flat in M^{n+k} when $k = 2$, then any imbedding of X^k into M^n is an ε-tame imbedding of X^k into M^{n+k}.

It follows immediately from Theorem 1 that an imbedding of X^k into M^n ($2k + 2 \leq n$) cannot fail to be locally tame at precisely one point. Applying this to a polyhedral neighborhood of a point of X^k, one obtains

THEOREM 10. If S is the set of points at which an imbedding f of X^k into M^n ($2k + 2 \leq n$) fails to be locally tame, then S contains no isolated points and must, therefore, be uncountable or empty.

If M and N are topological manifolds, of dimensions m and n and if f is an imbedding of M into N, then f is said to be <u>locally flat</u> at the point $x \in M$ if there is a neighborhood U of $f(x)$ in N such that the pair $(U, U \cap f(M))$ is homeomorphic to the pair (E^n, E^m). An imbedding f of a k-cell D into E^n is said to be <u>flat</u> if there is a homeomorphism of E^n onto itself carrying $f(D)$ onto the standard k-cell in the k-plane E^k in E^n. Similarly, f is locally flat at $x \in E$ if x lies in the interior (relative to D) of a k-cell D' in D on which f is flat. If $2k + 2 \leq n$, then an imbedding of D in E^n is flat if, and only if, it is tame.

Thus Theorem 1 implies

LEMMA 11. If f is an imbedding of the k-cell D into E^n ($2k + 2 \leq n$) such that $f|D - f^{-1}(E^\ell)$ ($2\ell + 2 \leq n$) is locally flat, then f is flat.

Cantrell and Edwards [4] have shown that there exists no "almost locally flat" imbedding (that is, an imbedding which is locally flat except at a countable number of points) of a topological m-manifold into a topological n-manifold whenever $2m + 2 \leq n$. Using Lemma 11, one can show that the following more general situation exists.

THEOREM 12. Suppose that M, N, and Q are (not necessarily triangulable) topological manifolds of dimensions m, n and q, respectively, with $m, q \leq \frac{n}{2} - 1$, such that Q is a locally flat submanifold of N. Let f be an imbedding of M into N such that $f|M - f^{-1}(Q)$ is locally flat. Then f is locally flat.

REFERENCES

[1] R. H. Bing and J. M. Kister, "Taming complexes in hyperplanes," Duke Math. J., vol. 31 (1964) pp. 491-511.

[2] M. Brown, "Locally flat embeddings of topological manifolds," Annals of Math., vol. 75 (1962) pp. 331-341.

[3] J. C. Cantrell, "n-frames in Euclidean k-space," Proc. Amer. Math. Soc., vol. 15 (1964) pp. 574-578.

[4] J. C. Cantrell and C. H. Edwards, Jr., "Almost locally flat imbeddings of manifolds," Michigan Math. J., vol. 12 (1965) pp. 217-223.

[5] J. Dugundji, "An extension of Tietze's theorem, " Pacific J. of Math., vol. 7 (1951) pp. 353-357.

[6] C. H. Edwards, Jr., "Taming 2-complexes in high-dimensional manifolds," Duke Math. J., (to appear).

[7] H. Gluck, "Embeddings in the trivial range Bull. Amer. Math. Soc., vol. 69 (1963) pp. 824-831.

[8] ______, "Embeddings in the trivial range (complete," Annals of Math., vol. 81 (1965) pp. 195-210.

[9] C. A. Greathouse, "Locally flat, locally tame and tame embeddings," Bull, Amer. Math. Soc., vol. 69 (1963) pp. 820-823.

[10] V. K. A. M. Gugenheim, "Piecewise linear isotopy and embeddings of elements and spheres I, II," Proc. London Math. Soc. (3), vol. 3 (1953) pp. 29-53, 129-152.

[11] T. Homma, "On the imbeddings of polyhedra in manifolds," Yokohama Math. J., vol. 10 (1962) pp. 5-10.

[12] V. L. Klee, Jr., "Some topological properties of convex sets," Trans. Amer. Math. Soc., vol. 78 (1955) pp. 30-45.

[13] E. C. Zeeman, "Unknotting combinatorial balls," Annals of Math., vol. 78 (1963) pp. 501-526.

SOME NICE EMBEDDINGS IN THE TRIVIAL RANGE

by

Jerome Dancis*

1. Introduction.

This paper is concerned with criteria for determining when k-complexes which are embedded in combinatorial n-manifolds, $2k + 2 \leq n$, (the trivial range of dimensions), are tame.

Let M be a combinatorial n-manifold, K a finite k-complex, $2k + 2 \leq n$, and f a topological embedding of K into M. Also let $f(K) = A + B$ where A is locally tame and B is compact. Throughout this paper M, K, k, n, f, A and B will have the properties listed above.

Tatsue Homma [9] has shown that if M is compact, $f(K) = B$ is p.w. ℓ embedded in the interior of a compact, combinatorial n-manifold M_1, where M_1 is topologically embedded in M, then $f(K) = B$ is tame. Also if $\varepsilon > 0$ is given, then there is an ε-homeomorphism h of M onto itself such that $h(B)$ is a subpolyhedron of M and h is the identity outside an ε-neighborhood of B.

Gluck (Th. 9.1 of [6] and Th. 4.1 of [7]) has improved Homma's result by showing that the above result holds when M and M_1 are possibly non-compact combinatorial n-manifolds. Also, if K_1 is a subcomplex of K and $f|K_1$ is a piecewise linear then one can find an ε-homeomorphism h of M onto itself such that $h|f(K_1) = 1$ and $h(B)$ is a polyhedron.

Gluck [6] and [7] and Greathouse [8] use the above to prove

Theorem 1: If $f(K) = A$ then A is tame, i.e., locally tame k-complexes are tame in combinatorial n-manifolds, $2k+2 \leq n$.

Bing and Kister [1] have shown that when $M = E^n$, $f(K) = B$ and B is contained in an (n-k)-hyperplane of E^n, then $f(K) = B$ is tame.

John Cobb (unpublished) has shown the following: Let P be a complex, Q a subcomplex and N a combinatorial n-manifold with boundary. Let $g: P \to N$ be an embedding such that $g|Q$ is p.w.ℓ., $g(P-Q)$ is locally tame, $g(P-Q) \subset \text{Int } N$ and $\dim (P-Q) = k$, $2k + 2 \leq n$. If $\varepsilon > 0$ is given, then there exists an ε-push T of $g(P-Q)$, such that $T \circ g$ is p.w.ℓ. and the isotopy of the push does not move any point of $g(Q)$.

* Work on this paper was partially supported by the National Science Foundation.

In their theorems, Gluck and Bing-Kister have shown that $f(K)$ may be moved onto a subpolyhedron of the space by an ε-push of $f(K)$.

This paper is a proof of:

Theorem 2: A necessary and sufficient condition that a k-complex K, which is a closed subset of a combinatorial n-manifold (without-boundary) $n \geq 2k+2$, be tame in M is that K lie in the union of a countable number of locally-tame (n-k) simplices in M. (Note $n-k \geq k+2$)

We shall establish two special cases of Theorem 2. The first special case (Theorem 3) and Theorem 5 will be used to establish the second special case (Theorem 10). Corollary 2 of Theorem 10 will then be used to establish Theorem 2. Corollary 2 of Theorem 10 will then be used to establish Theorem 2. Corollary 2 of Theorem 10 is dependent upon some of the "immediate consequences" (in section 5) of the first special case and Theorem 5.

2. Definitions

A complex L, topologically embedded in a combinatorial manifold M, is a polyhedron if the injection map $i: L \to M$ is piecewise-linear.

A complex L, topologically embedded in a combinatorial manifold M, is tame if there is a homeomorphism h of M onto itself such that $h(L)$ is a polyhedron.

Let L be a complex, embedded in a combinatorial manifold M. An open subset U of L is locally polyhedral if for each point x in U, there is a compact neighborhood C of x in U such that C is a subpolyhedron of both L and M. An open subset V of L is locally tame if for each point x in V, there is a compact neighborhood C of x in V and an open neighborhood N of C in M such that C is a subpolyhedron of L and C is tame in N.

An ambient ε-isotopy of M is a map $H: M \times I \to M$, where M is a topological space such that if for each $t \in [0, 1]$,

$$h_t(x) = H(x, t)$$

then h_t is a homeomorphism of M onto itself; $h = 1$, and $d(h_t(x), x) < \varepsilon$, for each $x \in M$, $t \in I$.

An ε-push P of A is an ambient ε-isotopy of M such that if $d(x, A) \geq \varepsilon$

$$h_t(x) = x, \quad \text{for all } t \in I.$$

P is said to push A onto $h_1(A)$, $P(x) = h_1(x)$.

3. Motivation

The proof of Theorem 3 was motivated by the following problem set which was assigned by R. H. Bing.

(NOTE: A stable homeomorphism is a homeomorphism which is the composition of a finite number of hoemomorphisms each of which is the identity on some open set.)

Problem 1. Suppose A is an arc in E^n $(n \geq 4)$ which is the sum of a countable number of segments and points. Then there is a line L in E^n such that no line parallel to L intersects A in two points.

Problem 2. Generalize 1 to suppose that A is a set in E^n which is the sum of a countable number of simplexes of dimension less than k.

Problem 3. If A, L arc as in 1, there is an $(n-1)$-plane P and a stable homeomorphism $h\colon E^n \to E^n$ such that $h(A) \subset P$ and h is invariant on each line parallel to L.

Problem 4. If A is an arc in the plane E^{n-1} of E^n, then there is a stable homeomorphism $h\colon E^n \to E^n$ such that no two points of $h(A)$ have the same last coordinate.

Problem 5. Each arc in E^{n-1} is stably flat in E^n.

Problem 6. Generalize 5 to topological k-complexes in m-planes of E^n.

(Klee proved such an extension.)

Using the methods suggested by the above problems one can easily provide proofs for the following problems too.

Problem 7. Suppose that A_1 is a compact set contained in the hyperplane E^{n-1} of E^n. Let $g\colon A_1 \to E^n(g(A_1) = A_2)$, be a homeomorphism. Then there is a stable homeomorphism $h\colon E^n \to E^n$ such that the last coordinates of a point x in $h(A_1)$ has the same value as the last coordinate of $g \cdot h^{-1}(x)$ in A_2.

Problem 8. Suppose A is an arc in E^n, $n \geq 4$, which is contained in a countable number of k-dimensional hyperplanes, $2k + 2 \leq n$. Then there is a stable homeomorphism of E^n onto itself such that $h(A)$ is a canonical interval I.

This method of showing that an arc A in E^n, $n \geq 4$, which is the sum of a countable number of straight line segments and points, is a tame arc, is a modification of the original method of proof due to Cantrell and Edwards [3].

4. The first special case.

Suppose τ_γ is a subset of a simplex σ_γ, then we shall say that $f\colon \tau_\gamma \to M$ is linear w.r.t. τ_γ if $f|\tau_\gamma$ has a linear extension $(\bar{f})_\gamma\colon \sigma_\gamma \to M$.

Let $f\colon K \to M$. We shall say that f is countably linear if K may be written as the countable union of subsets τ_γ, $\gamma = 1,2,\ldots$, where each τ_γ is contained in a simplex σ_γ of K such that $f|\tau_\gamma$ is linear w.r.t. τ_γ for each γ.

THEOREM 3. Suppose $f\colon K \to E^n$ where K is a finite k-complex, $2k+2 \leq n$ and f is a countably linear embedding. Then f is a tame embedding of K into E^n. Also if $\varepsilon > 0$ is given, then there is an ε-push p of $f(K)$ such that $p\cdot f\colon K \to E^n$ is p.w.l. Furthermore, under the isotopy of the push, each point of E^n may be made to move along a polygonal path, having length less than ε.

Proof: A rewording of Lemmas 3.3, 3.4 and 3.5 of the Bing-Kister paper [1] yields:

LEMMA 1 (Bing and Kister): Let π be the natural projection of $E^n = E^k \times E^{n-k}$ onto E^k. Let K be a finite k-complex, $f_k\colon K \to E^n$ an embedding, $2k + 2 \leq n$. Let i be a p.w.l. embedding of K into E^n such that $\pi\cdot i\colon K \to E^k$ is 1-1 on each simplex of K. If $\pi\cdot i(x) = \pi\cdot f_k(x)$, for each $x \in K$, and if $\varepsilon > 0$ is given, then there is an $\varepsilon/3k$-push p of $f_k(K)$ sending $f_k(K)$ onto a polyhedron such that, under the push, each point of E^n moves along a polygonal path of length less than $\varepsilon/3k$.

We obtain a p.w.l. injection $i\colon K \to E^n$ such that $\pi\cdot i\colon K \to E^k$ is 1-1 on each simplex of K by taking a piecewise-linear, $\varepsilon/3k$-approximation of f such that the vertices of the approximation are in general position and such that the images under π, of the vertices of each simplex of the approximation, are in general position in E^k.

Hence, thanks to the Bing-Kister lemma, the proof of Theorem 3 is reduced to finding $\varepsilon/2$-push P of $f(K)$ such that $\pi\cdot i(x) = \pi\cdot P\cdot f(x)$, for each $x \in K$. That is we must give $f(K)$ a push P which will make the first k-coordinates of $P\cdot f(x)$ agree with those of $i(x)$ for each $x \in K$. We will adjust the coordinates, one coordinate at a time (in order) and this of course we will do by induction.

In order to simplify the machinery we shall assume that the original triangulations of K and E^n had small mesh and that $i\colon K \to E^n$ is simplicial. We will identify K with its image under i.

Let $\pi_i\colon E^n = E^i \times E^{n-i} \to E^i$, $(i = 1, 2, \ldots, k)$ be the natural projection map. Note: $\pi_k = \pi$.

Theorem 3 will be established once we have verified the following.

Inductive claim: There is a set of homeomorphisms

$$\{f_0, f_1, \ldots, f_k;\ f_i\colon K \to E^n\}$$

and a set of $\varepsilon/2k$-pushes

$$\{P_1, \ldots, P_k;\ P_{i+1} \text{ is a push of } f_i(K) \text{ in } E^n\}$$

such that:

1. $f_0 = f$ and $f_{i+1} = P_{i+1} \circ f_i$;
2. $\pi_i f_i(x) = \pi_i(x)$, for each $x \in K$, and
3. f_i is countably linear with $(\bar{f}_i)_a$ denoting the linear extension of $f_i|_{\tau_a}$ defined in σ_a, and $\pi_i(\bar{f}_i)_a(x) = \pi_i(x)$, for each $x \in \sigma_a$, $a = 1, 2, \ldots$.

Proof of inductive claim: Assume that f_i and all of the $(\bar{f}_i)_a$, $a = 1,2, \ldots$ are known and satisfy conditions 2 and 3 above. The existence of an appropriate P_{i+1} is established by the two lemmas that follow.

Keeping in mind the following picture may help the reader juggle dimensional indices.

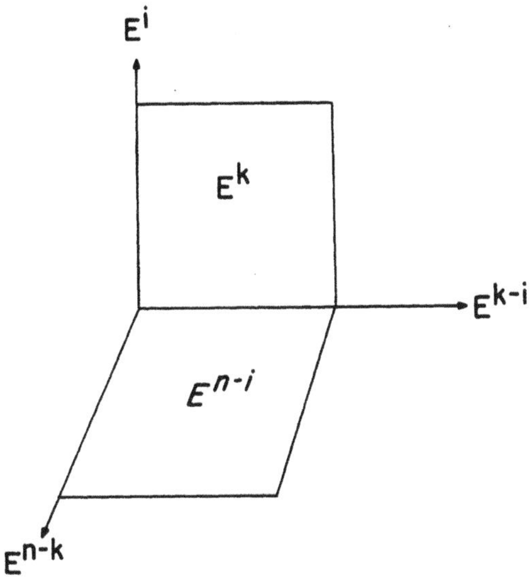

Fig. 1

In the figure, E^{n-k}, E^{k-i} and E^i are analogous to the X, Y and Z axis, respectively; E^{n-i} and E^k are analogous to the XY and YZ planes respectively.

LEMMA 2. Suppose that $f_i(K)$ is contained in the sum of a countable number of simplices σ'_γ, each of dimension $\leq k$, such that $\dim(\pi_i(\sigma'_\gamma)) \geq i - (k - \dim \sigma'_\gamma)$. Then there is a line L_{i+1} such that:

1. If L is any line, parallel to L_{i+1}, then $L \cap (\bigcup_{\gamma=1}^{\infty} \sigma')$ and
L $L \cap f_i(K)$ are each sets that contain at most one point;
2. $L_{i+1} \subset E^{n-i}$;
3. L_{i+1} goes through the origin, and the angle between L_{i+1} and the $(i+1)^{st}$ coordinate axis is arbitrarily small, and furthermore
4. We can coordinate L_{i+1} (and hence all lines parallel to it) in a continuous manner so that if $x \in L$ and L is parallel to L_{i+1}, then the coordinate value of x shall be the value of the $(i+1)^{st}$ coordinate of x.

Remark. If we let $\sigma_a' = (\bar{f}_i)_a(\sigma_a)$, $\dim \sigma_a' = r \leq k$, then the induction hypothesis says that $\pi_i(f_i)_a(x) = \pi_i(x)$, for all $x \in \sigma_a$, i.e., $\pi_i(\sigma_a') = \pi_i\sigma_a$. Now π is 1-1 on σ_a and hence $\dim(\pi_i(\sigma_a)) \geq r - (k-i) = i - (k-r)$.

Therefore $\dim(\pi_i(\sigma_a')) \geq i - (k-r)$ and the hypothesis of Lemma 2 is satisfied.

Proof of Lemma 2: Let $\rho \in \sigma_a'$ and suppose that ρ is the origin. This supposition is for convenience and is temporary. Now since $\dim(\pi_i(\sigma_a')) \geq i - (k-r)$ there is a linearly independent set of vectors $\{v_1, \ldots, v_{i-(k-r)}\}$ such that:

1. each vector is contained in σ_a, and
2. $< v_1, \ldots, v_{i-(k-r)} > \cap E^{n-i}$ = the origin.

Notation: $< A >$ means the hyperplane spanned by the elements of A.

Let σ_b' be an s-simplex, $s \leq k$. Then the two properties just listed imply that

$$\dim(< \sigma_a', \sigma_b' > \cap E^{n-i}) \leq r + s + 1 - [i - (k-r)] \quad .$$

Now

$$r + s + 1 - [i - (k-r)] = k + s + 1 - i$$

and since $k \geq s$, $n \geq 2k + 2$, we see that

$$\dim(< \sigma_a', \sigma_b' > \cap E^{n-i}) < n-i \quad .$$

Now the lines in a hyperplane H, if H intersects E^{n-i}, which are parallel to lines in E^{n-i} are also parallel to lines in $H \cap E^{n-i}$. Therefore the subset of the vector set

$$\{v: \text{ v goes from a point } a \text{ to a point } b, \text{ for all } a, b \in \sigma_a' \cup \sigma_b'\}$$

which lie in the vector space E^{n-i} span a vector subspace H_{ab} (a hyperplane through the origin) whose dimension is strictly less than n-i.

Now since $f_i(K)$ is contained in the countable union of simplices $\{\sigma_\gamma'\}$, where $\dim \sigma_\gamma' \leq k$, it follows that the subset of the vector set

$$\{v: \text{ v goes from a point } a \text{ to a point } b, \text{ for all } a, b \in \cup_{\gamma=1}^{\infty} \sigma_\gamma'\}$$

which lie in E^{n-i}, lie in $\cup_{\gamma,k=1}^{\infty} H_{\gamma,k}$ which does not contain any open set in E^{n-i}. (Courtesy of the Baire Category Theorem.)

Hence we can find a line L_{i+1} which satisfies conditions 1 to 4, and the lemma is established.

Now we can find an $\varepsilon/2k$-push P_{i+1} which will adjust the $(i+1)^{st}$ coordinate of $f_i(K)$ without disturbing the first i coordinates, i.e.:

LEMMA 3. Given the line L_{i+1} of the preceding lemma, and the induction hypothesis, there is an $\varepsilon/2k$-push P_{i+1} where:

1. $P_{i+1}(L) = L$, for all lines L parallel to L_{i+1};
2. $\pi_{i+1}P_{i+1}f_i(x) = \pi_{i+1}(x)$, for all $x \in X$, and
3. $f_{i+1} = P_{i+1}f_i$: $K \to E^n$ is countably linear and $\pi_{i+1}(\bar{f}_{i+1})_a(x) = \pi_{i+1}(x)$, for each $x \in \sigma_a$, $a = 1,2,\ldots$.

Proof: Let E^{n-1} be a hyperplane which meets L_{i+1} in exactly one point. Let ρ: $E^n \to E^{n-1}$ be the natural projection map which sends each line L, parallel to L_{i+1}, onto the point $L \cap E^{n-1}$.

Let h be a solution to problem 7 for $A_1 = \rho f_i(K)$, $A_2 = f_i(K)$, $g^{-1} = \rho|A_2$ such that ρh is the identity on E^{n-1}.

Now we define a new map

$$r\colon A_1 \to E^1$$

such that

$$r(x) = (i+1)^{st} \text{ coordinate of } f_i^{-1} g(x)$$

$$- \text{ the } L_{i+1} \text{ coordinate of } g(x),\ x \in A_1 = \rho f(K)$$

Next extend r to r': $E^{n-1} \to E^1$ such that either

$$|r'(x)| + d(h(x), A_2) < \frac{\varepsilon}{2k} ,$$

or

$$r'(x) = 0.$$

Now we may define P_{i+1} on $h(E^{n-1})$ as follows:

$$P_{i+1}(x) = x + r'p(x)\bar{u}, \quad x \in h(E^{n-1}) ,$$

where $\bar{u}$ is a unit vector in the positive L_{i+1}-direction. It is evident that the $(i+1)^{st}$ coordinate of y and of $P_{i+1}\,f_i(y)$, $y \in K$ are identical.

We note that since $\pi_i f_i(x) = \pi_i(x)$, $x \in K$ by the induction hypothesis and since L_{i+1} is orthogonal to E^i, being contained in E^{n-i}, condition 1 of this lemma implies that $\pi_i P_{i+1} f_i(x) = \pi_i(x)$, $x \in K$. Therefore condition two is satisfied.

Since each line L parallel to L_{i+1} intersects $h(E^{n-1})$ in exactly one point, one can easily extend $P_{i+1}|h(E^{n-1})$ to an $\varepsilon/2k$-push P_{i+1} such that condition 1 holds.

Condition 3. By the induction hypothesis $(\bar{f}_i)_m$: $\sigma_m \to E^n$, $m = 1, 2,\ldots,$ is a linear map. For each m, $m = 1,2,\ldots,$ we will obtain a linear homeomorphism

$$(\bar{P}_{i+1})_m\colon\ (\bar{f}_i)_m\,(\sigma_m) \;=\; \sigma'_m \to E^n$$

such that

$$(\bar{P}_{i+1})_m(\bar{f}_i)_m(x) \;=\; P_{i+1}f_i(x) \quad \text{for } x \in \tau_m \;.$$

Then we define $(\bar{f}_{i+1})_m = (\bar{P}_{i+1})_m\cdot(\bar{f}_i)_m$, and we see that $(\bar{f}_{i+1})_m$ is linear, since it is the composite of two linear maps. Thus f_{i+1} will be countably linear.

$(\bar{P}_{i+1})_m$ will be the composite of the following two linear homeomorphisms, each of which, like P_{i+1}, moves points only along the lines which are parallel to L_{i+1}. (Thus $\pi_i(\bar{f}_{i+1})_m(x) = \pi_i(\bar{f}_i)_m(x) = \pi_i(x)$, $x \in \sigma_m$). Since each line parallel to L_{i+1} intersects σ'_m down into the hyperplane E^{n-1}. Next we raise up $p(\sigma'_m)$ by matching the L_{i+1} coordinate value and hence the $(i+1)^{st}$ coordinate value of $p(x)$ with the $(i+1)^{st}$ coordinate value of x, for each $x \in \sigma_m$. This and the already mentioned fact that $\pi_i(\bar{f}_{i+1})_m(x) = \pi_i(x)$ implies that $\pi_{i+1}(\bar{f}_{i+1})_m(x) = \pi_{i+1}(x)$, $x \in \sigma_m$.

This completes the proofs of condition 3, Lemma 3, and the induction claim.

We note that under each push P_i, each point moves along a straight line. This completes the proof of Theorem 3.

Since locally tame complexes are tame (Th. 1), Theorem 3 may be generalized to:

THEOREM 4. Suppose that $f(K) = B$, where f is countably linear. Then B is tame.

5. Consequences

The next useful theorem follows directly from Theorems 9.1 and 10.1 of Gluck's paper [7]. This theorem may also be established by adapting the original methods of Tatsuo Homma. This theorem appears as Lemma 2.3 in [4], also see [5].

THEOREM 5. Let $f(K) = A + B$, B compact. Let $\varepsilon > 0$ be given. Then there is an ε-push P of $A - B$ such that $P(A-B)$ is locally polyhedral and P is the identity on B.

The remainder of this article is devoted to consequences of the main theorem and this theorem.

First, I shall provide an alternate proof to a theorem of Bryant [2], and Cobb (unpublished).

THEOREM 6. (Bryant, Cobb) Let $f(K) = A + B$, where B is a tame subcomplex. Then $f(K)$ is tame.

Proof:. Step 1. Send B into a polyhedron using a space homeomorphism h.
Step 2. Use Theorem 5 to send $A - B$ onto a locally polyhedral set via a space homeomorphism which leaves $h(B)$ fixed.
Step 3. Apply Theorems 3 and 1.

Theorem 6 immediately implies (via induction):

THEOREM 7. A k-complex (embedded in M) is tame if each of its simplices is tame.

THEOREM 8. A k-complex (embedded in M) is tame if the interior of each of its simplices is locally tame.

The next theorem has also been established by Bryant [2], and Cobb (unpublished) with somewhat different methods.

THEOREM 9. Let $f(K) = A + B$, where B is a countable set. Then $f(K)$ is tame.

The proof is easily obtained by using steps 2 and 3 of the proof of Theorem 6. Cantrell and Edwards have obtained this result under the additional hypothesis that K be a k-manifold [4].

6. The second special case.

In this section, we shall establish the second special case of Theorem 2, namely:

THEOREM 10. Let $f\colon I^k \to E^n$, $n \geq 2k+2$, be an embedding. Suppose $f(I^k) = A + B$ where A is locally-tame and B is contained in an $(n-k)$-dimensional polyhedron P. Then $f(I^k)$ is tame.

Proof: Without loss of generality we may assume that the set A contains all the open sets U_α such that U_α is locally-tame and that $B = f(I^k) - A$.

We shall show that if B is contained in an i-dimensional polyhedron P_i, then b is contained in an (i-1)-dimensional polyhedron P_{i-1}, $i = n-k, \dots, 2, 1, 0$.

Suppose that σ is an i-simplex of a triangulation of P_i, $p \in f^{-1}(B)$ and $f(p) \subset \text{Int}\,\sigma$. Then there is a canonical k-disk D in I^k such that $p \in \text{Int}\,D$ (with respect to I^k) and

$$D \subset f^{-1}(A \cup \text{Int}\,\sigma).$$

We now consider E^n as $E^{n-i} \times E^i$ where E^i is the hyperplane which contains σ. We note that $n-i \geq k$. Place a canonical copy of D in E^{n-i}. We now use Klee's method [10] or, equivalently, the methods of Problems 7 and 5 in order to find a homeomorphism h_1 of E^n onto itself which sends $f(D) \cap B$ onto the canonical copy of $D \cap f^{-1}(B)$ in E^{n-i}. Thus $h_1 f|D \cap f^{-1}(B)$ is linear with respect to $D \cap f^{-1}(B)$.

We now apply Theorem 5 where A, B and f of Theorem 5 correspond to $f(D) \cap A$, $f(D) \cap B$ and $f|D$ of this theorem, in order to obtain a second homeomorphism h_2 of E^n onto itself such that

$$h_2 h_1 f|D \cap f^{-1}(A) \text{ is locally piecewise-linear}$$

and

$$h_2 | h_1 f(D \cap f^{-1}(B)) = 1.$$

Thus $h_2 h_1 f|D$ is countably linear. Therefore Theorem 3 is applicable and we find that $f(D)$ is tame and hence $f(\text{Int}\,D)$ is locally-tame.

Therefore B may not intersect the interior of any i-simplex of a triangulation of P_i. Hence B is contained in an (i-1)-dimensional polyhedron P_{i-1} which is the (i-1)-skeleton of P_i.

The hypothesis of this theorem says that B is contained in a (n-k)-dimensional polyhedron. This and the inductive argument just given imply that $B = \varphi$ and hence $f(I^k)$ is locally-tame.

Finally we use Theorem 1 and then Theorem 10 is established.

COROLLARY 1. Let $f\colon I^k \to M$ be an embedding where M is a combinatorial n-manifold, $n \geq 2k+2$. Suppose $f(I^k) = A + B$ where A is locally-tame and B is contained in a (n-k)-dimensional polyhedron P or a tame (n-k)-complex or a locally-tame (n-k)-complex. Then $f(I^k)$ is locally-tame.

This corollary and Theorems 8 and 1 yield:

COROLLARY 2. Let f be an embedding of a k-complex K into a combinatorial n-manifold $n \geq 2k+2$. Suppose $f(K) = A + B$ where A is locally-tame and B is contained in a locally-tame (n-k)-complex. Then $f(K)$ is tame.

John Bryant [2] has established Corollary 2 to Theorem 10 when B is contained in a tame k_1-complex, $n \geq 2k_1+2$ instead of a locally-tame (n-k)-complex. Note: $(n-k) \geq k_1 + 2$.

COROLLARY 3. Let f be an embedding of a k-complex K into a topological n-manifold, $n \geq 2k+2$. Suppose that U is an open subset of $f(K)$ such that U may be written as $U = A + B$ where A is open, A is locally-tame and B is contained in a locally-tame (n-k)-complex. Then U is locally tame.

7. Conclusion.[†]

In this section, we shall establish Theorem 2.

Proof of Theorem 2: Let A be the open subset of $f(K)$ which is maximal with respect to the property A is locally-tame, i.e., A is the sum of all open subsets U of $f(K)$ such that U is locally-tame. Let X_j, $j = 1, 2, \ldots$ be a countable collection of locally-tame (n-k)-simplices in M whose union contains $f(K)$. Let $B = f(K) - A$, and let $Y_j = B \cap X_j$. Then B is compact, $B = \cup_{j=1}^{\infty} Y_j$, each Y_j is compact and no point of B is a member of an open set U such that U is locally-tame.

Now B is a complete metric space and therefore we may use the Baire Category Theorem to obtain a Y_j which contains an open subset O of B. If p is a point of O, then p has a compact neighborhood N_p in $f(K)$ such that

$$N_p \subset O \cup A$$

and N_p is a complex which is piecewise-linearly embedded in $f(K)$. Hence $A \cap N_p$ is locally tame with respect to the triangulation of N_p. Also

$$O \subset Y_j \subset X_j$$

and X_j is a tame (n-k)-simplex.

Hence $N_p - A = O \cap N_p$ is contained in a tame (n-k)-simplex.

Therefore we may apply Corollary 2 of Theorem 10 where the K, A, B and f of the corollary correspond to N_p, $A \cap N_p$, $N_p - A = O \cap N_p$ and $f|f^{-1}(N_p)$ of this theorem.

Therefore N_p is tame and since the triangulation of N_p is compatible with the triangulation of $f(K)$, the Int N_p is locally tame with respect to the triangulation of $f(K)$.

Now $p \in \text{Int } N_p$ and hence $p \in A$. But we had assumed

$$p \in O \subset Y_j \subset B = f(K) - A.$$

That is, we had assumed that A contained all the points p which had neighborhoods N_p such that N_p is locally tame. Thus, we have obtained the desired contradiction.

[†] The part of the proof presented in this section is due to R. H. Bing.

Therefore $B = \varphi$ and $f(K)$ is locally-tame. Apply Theorem 1 and then Theorem 2 is established.

John Bryant has independently discovered that he can use the method described in this section in order to generalize his main theorem [2]. Thus he can show that $f(K)$ is tame if it is contained in the countable union of tame k_1-complexes in M when $n \geq 2k_1+2$. Note: $n-k \geq k_1+2$.

REFERENCES

[1] R. H. Bing and J. M. Kister, "Taming complexes in hyperplanes," Duke Math. J., vol. 31 (1964) pp. 491-511.

[2] J. L. Bryant, "Taming Polyhedra in the Trivial Range," preceding paper in this book.

[3] J. C. Cantrell and C. H. Edwards, Jr., "Almost locally polyhedral curves in Euclidean n-space," Trans. Amer. Math. Soc., vol. 107 (1963) pp. 451-457.

[4] ___________, "Almost locally flat imbeddings of manifolds," Michigan Math. J., vol. 12 (1965) pp. 217-223.

[5] C. H. Edwards, Jr., "Taming 2-complexes in high-dimensional manifolds," Duke Math. J., (to appear).

[6] H. Gluck, "Embeddings in the trivial range," Bull. Amer. Math. Soc., vol. 69 (1963) pp. 924-831.

[7] ______, "Embeddings in the trivial range (complete)," Annals of Math., vol. 81 (1965) pp. 195-210.

[8] C. A. Greathouse, "Locally flat, locally tame and tame embeddings," Bull. Amer. Math. Soc., vol. 69 (1963) pp. 820-823.

[9] T. Homma, "On the imbeddings of polyhedra in manifolds," Yokohama Math. J., vol. 10 (1962) pp. 5-10.

[10] V. L. Klee, Jr., "Some topological properties of convex sets," Trans. Amer. Math. Soc., vol. 78 (1955) pp. 30-45.

APPROXIMATIONS AND ISOTOPIES IN THE TRIVIAL RANGE

by

Jerome Dancis

1. Introduction and statement of results

In this paper we will generalize some of the known results about embeddings of k-complexes into combinatorial n-manifolds, $n \geq 2k + 2$ (the "trivial" range of dimensions). Mainly, we shall show that in many cases the conditions about "global" triangulations of complexes and manifolds may be replaced by conditions about "local" triangulations of sets and manifolds. The main results of this paper are:

THEOREM 1 (Locally-flat approximations). Let g be a map of a compact, topological k-manifold-with-boundary M^k into a topological n-manifold-with-boundary M^n, $n \geq 2k + 2$. Then given an $\varepsilon > 0$, there exists a homeomorphism h of M^k into M^n such that $d(g, h) \leq \varepsilon$ and $h(M^k)$ is locally flat.

COROLLARY 1.1 (Locally-flat embeddings). Every compact, topological k-manifold-with-boundary may be embedded in E^{2k+2} as a locally-flat set.

THEOREM 2 (Locally-tame approximations I). Let g be a map of a k-complex K into a topological n-manifold-without-boundary M^n, $n \geq 2k + 2$. Then given an $\varepsilon > 0$, there is a (strongly)-locally-tame embedding h of K into M^n such that $d(g, h) \leq \varepsilon$.

THEOREM 3 (Locally-tame approximations II). Let L be locally a k-complex. Let g be a map of L into a topological n-manifold-without- boundary M, $n \geq 2k + 2$. Then given an $\varepsilon > 0$, there exists a homeomorphism h of L into M such that $d(g, h) \leq \varepsilon$ and $h(L)$ is (weakly) locally-tame.

THEOREM 4. Let L be locally k-complex. Suppose that f_0 and f_1 are two embeddings of L in E^n, $n \geq 2k + 2$ such that $f_0(L)$ and $f_1(L)$ are (weakly) locally-tame. Then there is a push P such that

$$P \cdot f_0 = f_1 \ .$$

COROLLARY 4.1. Suppose that f and g are two embeddings of a compact topological k-manifold-with-boundary M^k into E^n, $n \geq 2k + 2$, such that $f(M^k)$ and $g(M^k)$ are locally flat. Then there is a homeomorphism h of E^n onto itself such that $hf = g$ and h is the identity off a compact set.

We shall present a summary of some of the known results on tame k-complexes in combinatorial n-manifolds $n \geq 2k + 2$, in section 2.

In section 4 we shall show that a "weak" definition of locally-tame and the definition of locally flat are equivalent for k-manifolds-with-boundary, which are embedded in n-manifolds, $n \geq 2k + 2$, (Theorem 18), and that the "weak" and "strong" definitions of tame and locally-tame are equivalent for k-complexes in n-manifolds, $n \geq 2k + 2$, (Theorems 16 and 17).

We shall set the stage for proving Theorems 1, 2, 3, and 4 in section 5. Here, we shall also establish

LEMMA 5. Let f be a map of an r-complex K into a combinatorial n-manifold M. Let X be the finite union of tame s-complexes in M, $r + s < n$. If $\varepsilon > 0$ is given, then there is a piecewise linear map

$$g: K \to M$$

such that

$$g(K) \cap X = \emptyset ,$$

and

$$d(g, f) < \varepsilon .$$

In section 6 we shall establish Theorems 1, 2, and 3, and in section 7 we shall establish Theorem 4.

2. Background

V. K. A. M. Gugenheim (Theorem 5 [8]) proved:

THEOREM 6. Let K be a finite k-polyhedron which is piecewise-linearly embedded in E^n, $n \geq 2k + 2$, and let f be a piecewiselinear homeomorphism of K into E^n. Then there is a piecewiselinear homeomorphism h of E^n onto itself such that $h|K = f$.

As Homma observed, (also see p. 506 [1]), Gugenheim's proof of Theorem 6 induces the following, more general theorem:

THEOREM 7. Under the hypothesis of Theorem 6 if K^* is a subpolyhedron of K such that $f|K^* = 1$, then for any $\varepsilon > 0$, there is a piecewiselinear homeomorphism h of E^n onto itself such that

$$h|K = f;$$

$$d(h, 1) < d(f, 1) + \varepsilon,$$

and

$$h|E^n - U_\varepsilon\Big(\bigcup_{t \in I} f_t(K-K^*)\Big) = 1 ,$$

where $f_t(x) = (1-t)x + tf(x)$, $t \in [0,1]$.

Notation: $U_\varepsilon(A)$ means an ε-neighborhood of the set A.

Homma uses Theorem 7 to prove the following theorem (see the Corollary to Lemma 2 [9]) and his main theorem [9]:

THEOREM 8. Let f be a piecewiselinear homeomorphism of a finite k-polyhedron K which is in the interior of a compact combinatorial n-manifold with boundary M, into $\operatorname{Int} M$ such that f is homotopic to the identity and $n \geq 2k + 2$. Then f can be extended to a piecewiselinear homeomorphism h of M onto itself.

A special case of Homma's main theorem [9] is:

THEOREM 9. Let K be a finite k-complex, topologically embedded in a combinatorial n-manifold M, $n \geq 2k + 2$. If K is tame and if $\varepsilon > 0$ is given, then there is an ε-homeomorphism h of M onto itself such that $h|K$ is piecewise linear with respect to K and $h|M - U_\varepsilon(K) = 1$.

Homma's proofs of Theorems 8 and 9 may be combined to prove:

THEOREM 10. Under the hypothesis of Theorem 9, if K^* is a subcomplex of K, f is a homeomorphism of K into M such that $f(K)$ is tame, $f|K^* = 1$ and there is a homotopy

$$\{f_t | f_t\colon K \to M,\ f_t|K^* = 1,\ t \in [0, 1]\},\quad f_0 = 1 \text{ and } f_1 = f,$$

then there is a homeomorphism h of M onto itself such that

$$h|K = f$$

and

$$h|M - U_\varepsilon\Big(\bigcup_{t \in I} f_t(K-K^*)\Big) = 1 .$$

Bing and Kister (see section 5 [1]) have sharpened Gugenheim's result (Theorem 7) by showing that one may obtain an h which is ambient isotopic to the identity namely:

THEOREM 11. Under the hypothesis of Theorem 6 if K^* is a subcomplex of K such that $f|K^* = 1$, then for any $\varepsilon > d(f, 1)$, there is an ε-push P of $(K-K^*)$ such that

$$P|K = f,$$

P does not move any point of K^*, and each point of E^n is moved along a polygonal path of length no more than ε by the isotopy of the push. Furthermore, if $\delta > 0$ is given, then one may find an ε-push P such that

$$P|E^n - U_\delta\Big(\bigcup_{t \in I} f_t(K-K^*)\Big) = 1$$

where

$$f_t(x) = (1-t)x + tf(x), \quad x \in K, \quad t \in I = [0, 1].$$

Remark. In order to adapt Bing and Kister's proof for this more general case, one should reorder the vertices of K, so that if K^* has r vertices then $\{v_1, \ldots, v_r\} \subset K^*$. Now let $f(x, t) = f(x, 0)$ for all $x \in K^*$, $t \in [0, 1]$ and omit the first r-parts of h_t on pages 501, 503 [1], and then section 5 [1] will prove Theorem 11.

Suppose we used Bing and Kister's work (Theorem 11) instead of Gugenheim's result (Theorem 7) in the proofs of Homma's theorems [9] as Gluck does in [7]. Then we may establish the following two theorems which are similar to some of Gluck's results [7].

THEOREM 12. Suppose K is a k-complex topologically embedded in E^n, $n \geq 2k + 2$, and K^* is a subcomplex of K. Let $f\colon K \to E^n$ be an ε-homeomorphism such that $f|K^* = 1$. Suppose that K and $f(K)$ are tame. Then for any $\varepsilon_1 > \varepsilon$, there is an ε_1-push P on $K-K^*$ such that

$$P|K = f,$$

and P does not move any point of K^*.

THEOREM 13. Suppose K is a k-complex topologically embedded in a combinatorial n-manifold M, $n \geq 2k + 2$, and K^* is a subcomplex of K. Let

$$\{f_t | f_t\colon \; K \to M, \;\; f_t|K^* = 1, \;\; t \in [0, 1], \;\; f_0 = 1\}$$

be a homotopy. Suppose $f = f_1$ is a homeomorphism, K and $f(K)$ are tame and $\varepsilon > 0$ is given. Then there is an ambient isotopy $\{h_t\}$ such that

(i) $h_0 = 1$,

(ii) $h_1|K = f$,

(iii) $h_t|M - U_\varepsilon\Big(\bigcup_{t \in I} f_t(K-K^*)\Big) = 1$,

(iv) $h_t|K^* = 1$.

Gluck [5], [6] and Greathouse [7] use many of the results which are listed here in order to establish:

THEOREM 14. A k-complex, in a combinatorial n-manifold, $n \geq 2k + 2$, with a (strongly)-locally-tame embedding is tame.

3. Definitions

A subset of a complex K is a polyhedron in K or a subpolyhedron of K if it is the point set of some subcomplex of some subdivision of K.

An embedding f of a complex K into a combinatorial manifold-with-boundary M is (strongly) tame if there is a homeomorphism h of M onto itself such that $h \cdot f$ is piecewiselinear.

A subset K of a combinatorial manifold-with-boundary M is (weakly) tame if there is a homeomorphism h of M onto itself, such that $h(K)$ is a polyhedron in M.

We shall say that a complex embedded in a combinatorial manifold-with-boundary is tame if it has a (strongly) tame embedding.

Let f be an embedding of a complex K into an n-manifold-with-boundary M and let A be an open subset of K. We say that $f|A$ is (strongly) locally-tame if for each point $x \in A$, $f(x)$ has a compact neighborhood U_x in M and if there is a homeomorphism h_x sending U_x onto I^n such that $h_x f|f^{-1}(U_x)$ is piecewiselinear.

A subset U of a topological n-manifold-with-boundary M is (weakly) locally-tame, if each point $x \in U$ has a compact neighborhood N_x and if there is a homeomorphism h_x sending N_x onto I^n such that $h_x(U \cap N_x)$ is a subpolyhedron of I^n.

We note that the "strong" definitions used here are equivalent to the definitions used in the preceding paper [4].

A q-manifold-with-boundary M_1, topologically embedded in a topological manifold M, is locally flat in M if each point $x \in M_1$ has a neighborhood U in M such that the pair $(U, U \cap M_1)$ is homeomorphic to (E^n, E^q) or (E^n, E^q_+) where E^q_+ is a q-dimensional holf-plane in E^n.

A compact set L is locally a k-complex if each point $x \in L$ has a compact neighborhood which is homeomorphic to a k-complex. We note that every compact, topological k-manifold, every finite k-complex and every compact, (weakly)-locally-tame set of dimension k is locally a k-complex.

An ambient isotopy of a space M is a map $H\colon M \times I \to M$, such that if we set

$$h_t(x) = H(x, t),$$

then h_t is a homeomorphism of M onto itself, for each $t \in I$, and $h_0 = 1$.

We say that h_a is "(ambient) isotopic" to h_b, for $a, b \in I$.

An ambient ε-isotopy of M is an ambient isotopy which satisfies the additional condition:

$$d(h_t(x), x) \leq \varepsilon, \text{ for each } x \in M, \quad t \in I .$$

A push is an ambient isotopy of a space M such that for some compact proper subset A' of M

$$h_t(x) = x, \quad \text{for all } x \in M - A' \text{ and } t \in I,$$

and $h_0 = 1$. An ε-push P of A is an ambient ε-isotopy of M such that

$$h_t(x) = x, \quad \text{when } d(x, A) \geq \varepsilon, \quad t \in I$$

and $h_0 = 1$. We say that P "pushes" or "moves" A onto $h_1(A)$, and

$$\underline{P(x)} = h_1(x).$$

Also, "P does not move any point of a set B" means that $h_t(x) = x$, for all $x \in B$ and $t \in I$.

We define two rules of "composition" of pushes as follows. Suppose P, P' and P'' are three pushes which represent the isotopies $\{h_t, t \in I\}$, $\{h_t'\}$ and $\{h_t''\}$, respectively, on a space M, and h and h^* are homeomorphisms of M onto itself, then

(i) $P' = h^*Ph$ means $h_t' = h^*h_th$, $t \in I$, and

(ii) $P'' = P'\cdot P$ means $h_t'' = \begin{cases} h_{2t}, & t \in [0, \frac{1}{2}], \\ h_{2t-1}'\cdot h_1, & t \in [\frac{1}{2}, 1]. \end{cases}$

Notation: (i) $U_r(S) = \{x, d(x, S) < r\}$

(ii) Let $f, g\colon X \to Y$, then $d(f, g) = \operatorname{lub}_{x \in X} d(f(x), g(x))$.

(iii) pwl means piecewise linear.

Manifolds are manifolds-without-boundary, and all complexes are finite.

4. Equivalence of definitions

In this section we shall establish several equivalence relations among the various types of "nice" embeddings in the "trivial range."

We begin by stating a very useful lemma.

LEMMA 15. Let f be an embedding of a k-complex K into a topological n-manifold M, $n \geq 2k + 2$. Suppose U is an open subset of K such that $f(U)$ is contained in some (weakly) locally-tame set Y, $\dim Y \leq k$. Then $f|U$ is (strongly) locally-tame. Furthermore, if $K = U$ and M has a combinatorial triangulation, then f is (strongly) tame.

Lemma 15 is a special case of Corollarys 2 and 3 of Theorem 10 [4]; also it is essentially a corollary of Bryant's main theorem [3].

We now show the equivalence of the "strong" and "weak" definitions in the "trivial range" of dimensions.

THEOREM 16. Let K be a k-complex, M a combinatorial n-manifold, $n \geq 2k + 2$ and f an embedding of K into M. Then f is (strongly) tame iff $f(K)$ is (weakly) tame.

THEOREM 17. Let f be an embedding of a k-complex K into a topological n-manifold M, $n \geq 2k + 2$. Let U be an open subset of K. Then $f|U$ is (strongly) locally-tame (with respect to any triangulation of K), iff $f(U)$ is a (weakly)-locally-tame set.

<u>Proof of Theorems 16 and 17.</u>

(i) The "strong" definitions always imply the "weak" definitions.

(ii) That the "weak" definitions imply the "strong" definitions follows directly from Lemma 15.

Theorem 17 generalizes a result of Gluck (Theorem 6.1 of [5] or Theorem 11.1 of [6]) which says that if an embedding of a k-complex K into a combinatorial n-manifold, $n \geq 2k + 2$, is (strongly) locally-tame with respect to one triangulation of K then it is (strongly) locally-tame with respect to any triangulation of K. Theorems 17 and 14 have the following corollary:

COROLLARY 17.1. (Weakly) locally-tame k-complexes in combinatorial n-manifolds, $n \geq 2k + 2$, are tame.

The remainder of this section is about some relationships between local flatness and local tameness.

Let f be an embedding of a combinatorial q-manifold-with-boundary M^q into a topological n-manifold M^n. Greathouse [7] has used Morton Brown's bicollaring theorem [2] to show that f is (strongly) locally-tame if $f(M^q)$ is locally flat and $\partial M^q = \emptyset$ (for all $q \leq n$). One may combine Greathouse's method with an (unpublished) technique of R. H. Bing (for proving that locally-flat q-cells are flat in E^n, $q \leq n$) in order to show that Greathouse's result is true without the condition $\partial M^q = \emptyset$, i.e., for embeddings of combinatorial manifolds-with-boundary, locally flat implies (strongly) locally-tame.

THEOREM 18. A topological k-manifold-with-boundary M^k, embedded in a topological n-manifold M^n, $n \geq 2k + 2$, is locally flat iff it is (weakly) locally-tame.

<u>Proof</u>: (i) The definitions imply that locally flat always implies (weakly) locally-tame.

(ii) (Weakly) locally-tame $\Longrightarrow$ locally flat:

The definition of (weakly) locally-tame says that each point $x \in M^k$ has a compact neighborhood N_x in M^n and that there is a homeomorphism h_1 of N_x onto I^n such that $h_1(N_x \cap M^k)$ is a subpolyhedron Q of I^n (with $\dim Q = k$). The definition of manifold-with-boundary tells us that the point x has a compact neighborhood B^k in M^k such that $B^k \subset \text{Int } h_1^{-1}(Q)$ and there is a homeomorphism h_2 of B^k onto a k-cell I^k.

Therefore we may now use Lemma 15 to show that $h_1h_2^{-1}: I^k \to I^n$ is (strongly) tame. Therefore there is a homeomorphism h_3 of $\text{Int } I^n$ onto itself such that

$$h_3h_1h_2^{-1}: I^k \to \text{Int } I^n$$

is piecewise linear. We may now use Theorem 6 in order to find another homeomorphism h_4 of $\text{Int } I^n$ onto itself such that $h_4h_3h_1h_2^{-1}(I^k) = h_4h_3h_1(B^k)$ is a canonical k-cell I_1^k in $\text{Int } I^n$.

Thus, Theorem 18 is established.

5. The set-up

In this section, we will establish a "general position" lemma (Corollary 19.1) and a "canonical representation" for compact sets which are locally complexes. These results and Lemma 15 are the basis of the proofs of Theorems 1, 2, 3 and 4.

LEMMA 19. Let:

(i) K be a $(k+1)$-complex;

(ii) K' and K'' be subcomplexes of K;

(iii) $K' \cup K'' = K$;

(iv) $f: K \to E^n$, $n \geq 2k + 2$ be a map;

(v) $f|K'$ be a homeomorphism;

(vi) X be a subset of E^n which is contained in the finite union of tame k-complexes or a (weakly) locally-tame set of dimension $\leq k$;

(vii) $f(K') \cap X = \emptyset$;

(viii) $d(f(K' \cap K''), X) > 0$, and

(ix) $\varepsilon > 0$ be given.

Then there is a map $g: K \to E^n$ such that

(x) $g|K' = f|K'$;

(xi) $d(g, f) < \varepsilon$;

(xii) $g(K) \cap X = \emptyset$ and

(xiii) $d(g(K''), X) > 0$.

Note: (vii), (x), (xiii) and (iii) imply (xii).

COROLLARY 19.1. If K' is a k-complex and $f|K'$ is a piecewise-linear homeomorphism in Lemma 19, then we may find a g which is a piecewise-linear, general position map and satisfies (x)-(xiii) above.

Proof of Corollary 19.1: We obtain the desired map by taking a general-position, piecewise-linear approximation of the map obtained from Lemma 19 which agrees with f on K'.

Proof of Lemma 19: First, we note that every (weakly) locally-tame set is contained in a finite union of tame complexes and hence X is contained in the union of m tame k-complexes, for some integer m. This proof will be an induction on "m".

Case 1. $m = 1$, i.e., X is contained in a tame k-complex.

Step 1. Send the tame k-complex (which contains X) onto a polyhedron by a space homeomorphism h.

Step 2. There is a $\delta > 0$ such that if $p, q \in h[U_1(f(K))]$ and $d(p, q) < \delta$, then $d(h^{-1}(p), h^{-1}(q)) < \varepsilon$.

Step 3. Now we can find a map $g': K \to E^n$ such that g' is piecewise linear outside of a small neighborhood of K', $d(g'(K''), h(X)) > 0$, $d(h \cdot f, g') < \delta$ and $g'|K = h \cdot f|K'$.

Step 4. The desired map g is $h^{-1}g'$.

Case 2. Suppose Lemma 19 is true whenever X is contained in the union of $(m-1)$ tame k-complexes, and now we will use this supposition to prove that the lemma is true when X is contained in the union of m tame k-complexes.

Step 1. We may decompose X into two sets X_1 and X_2, where X_1 is contained in a tame k-complex and X_2 is contained in the union of $(m-1)$ tame k-complexes.

Step 2. We use Case 1 on X_1 in order to obtain a map

$$g_1: K \to E^n$$

such that

$$g_1|K' = f|K',$$

and

$$d(g_1, f) < \varepsilon/2.$$

Step 3. Let $\varepsilon_1 = d(g_1(K''), X_1)$.

Step 4. We note that our supposition enables us to use Lemma 19 on X_2 in order to obtain a map $g_2: K \to E^n$ such that

$$d(g_2(K''), X_2) > 0$$

and

$$d(g_2, g_1) < \min(\varepsilon/2, \varepsilon_1).$$

Step 5. A little checking shows that g_2 satisfies conditions (x)-(xiii).

This completes the proof of Lemma 19.

Remark. The proof of Lemma 5 is essentially the same as the proofs of Lemma 19 and Corollary 19.1 for the case $K' = \emptyset$.

<u>Canonical representations.</u> If L is locally a complex, then a canonical representation of L is a collection of compact subsets $\{L_j, K_j, K'_j, K''_j, j = 1, 2, \ldots, r\}$ of L such that (see Figure 1):

(0) $L_1 = K_1 = K''_1$, $K'_1 = \emptyset$ and $L_r = L$;

(1) K_j is a complex;

(2) $L = \cup_j \operatorname{Int} K_j$;

(3) K'_j and K''_j are subcomplexes of K_j, and $K'_j \cup K''_j = K_j$;

(4) $L_j \subset L_{j-1} \cup K_j$; $j > 1$;

(5) $\operatorname{Int} L_j \supset L - \cup_{m>j} \operatorname{Int} K_m$;

(6) $K'_j \subset \operatorname{Int} L_{j-1}$, $j > 1$;

(7) $K''_j \subset \cup_{m \geq j} \operatorname{Int} K_m \cup$

and

(8) $d(K''_j, L_j - K_j) > 0$, $j > 1$.

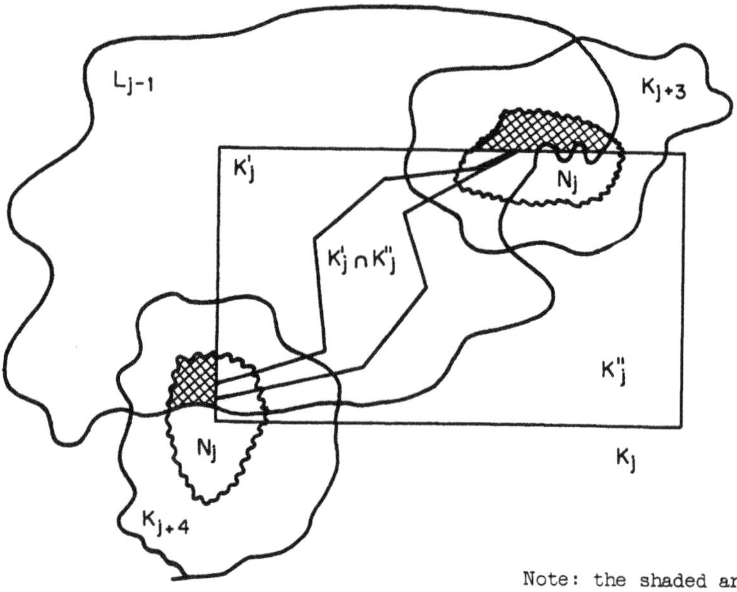

Note: the shaded area is $L_{j-1} - L_j$.

Figure 1

LEMMA 20. Every compact set L, which is locally a complex, has a canonical representation. Furthermore if $\{K_j, j = 1, 2, \ldots, r\}$ is a particular set of complexes such that $L = \cup_{j=1}^{r} \operatorname{Int} K_j$, then L has canonical representation whose collection of K_j's is the given set.

Proof: Suppose that L is a compact set which is locally a complex. Then the definition will provide a finite set of complexes $\{K_1, \ldots, K_r\}$ such that

$$L = \bigcup_j \operatorname{Int} K_j \ .$$

Therefore Conditions (1) and (2) are satisfied.

We begin by setting:

$$L_1 = K_1, \quad K_1' = \emptyset, \quad K_1'' = K_1 \quad \text{and} \quad L_r = L;$$

thus Condition (0) is satisfied.

We will construct the L_j's, K_j''s and K_j''s by induction.

We assume that L_{j-1} is known and satisfies Condition (5), namely:

$$\operatorname{Int} L_{j-1} \supset L - \bigcup_{m \geq j} \operatorname{Int} K_m \ .$$

Therefore

$$\partial L_{j-1} \subset \bigcup_{m \geq j} \operatorname{Int} K_m \ ,$$

and

$$L_{j-1} \cup \Big(\bigcup_{m \geq j} \operatorname{Int} K_m \Big) \supset L \supset K_j \ .$$

Thus K_j is contained in theuunion of two sets, one of which ($\bigcup_{m \geq j} \operatorname{Int} K_m$) is open and contains the boundary of the second. Therefore the complex K_j may be decomposed into two subcomplexes K_j' and K_j'' such that Conditions (3), (6) and (7) are satisfied.

Condition (7) implies that there is an open set N_j such that

$$K_j' \cap K_j'' \cap \partial K_j \subset N_j \subset \text{closure of } N_j \subset \bigcup_{m \geq j} \operatorname{Int} K_m \ .$$

Now we are ready to define L_j namely:

$$L_j = \text{Closure } (L_{j-1} - N_j) \cup K_j \ .$$

A brief checking of the above equation and inequalities (and Figure 1) shows that Conditions (4), (5) and (8) are satisfied.

We have shown that all the conditions of a canonical representation are satisfied and therefore Lemma 20 is established.

6. The approximations.

The bulk of this section is devoted to the Proof of Theorem 3. The compact set L has a canonical representation (see Lemma 20) such that K_j $(j = 1, 2, \ldots, r)$ is a k-complex and $g(K_j)$ has a euclidean neighborhood E_j^n in M. Let

$$\varepsilon_1 = \min \{\varepsilon, \ d[g(K_j), \ M - E_j^n], \ (j = 1, 2, \ldots, r)\}.$$

The desired approximation will be constructed by induction, namely; we will find a series of homeomorphisms:

$$\{h_1, \dots, h_r;\ h_j:\ L_j \to M\}$$

such that

(i) $h_j(L_j)$ is contained in a (weakly) locally-tame set of dimension $\leq k$, and

(ii) $d(h_j,\ g|L_j) < \varepsilon_1/3^{r-j}$.

<u>Note</u>: $d(h_{j-1}(K_j'),\ M-E_j^n) > \varepsilon_1 - \varepsilon_1/3^{r-j-1}$.

Let $h_1:\ K_1 = L_1 \to E_1^n \subset M$ be a piecewise-linear, homeomorphic $\varepsilon_1/3^{r-1}$-approximation of $g|K_1$. (Here h_1 is piecewise linear with respect to K_1 and E_1^n, and $d(h_1,\ g|K_1) < \varepsilon_1/3^{r-1}$ with respect to the metric on M.)

Assume that $h_{j-1}:\ L_{j-1} \to M$ is known (and satisfies the two conditions listed above). We note that $h_{j-1}(K_j')$ is contained in a (weakly) locally-tame subset of E_j^n and therefore we may use Lemma 15 to show that $h_{j-1}(K_j')$ is tame in E_j^n. We now use Homma's result (Theorem 9) in order to obtain a homeomorphism of E_j^n onto itself and hence a homeomorphism h of M onto itself such that

$$h \cdot h_{j-1}|K_j':\ K_j' \to E_j^n \subset M \text{ is piecewise linear}$$

and

$$d(h,\ 1) < \varepsilon_1/3^{r-j-1} .$$

Hence

$$d(g|L_{j-1},\ h \cdot h_{j-1}) < 2\varepsilon_1/3^{r-j-1} .$$

Extend $h \cdot h_{j-1}|K_j'$ to a piecewise-linear map

$$h_j':\ K_j \to E_j^n \subset M$$

such that

$$d(g|K_j,\ h_j') < 2\varepsilon_1/3^{r-j-1} ,$$

We note that Condition (8) of a Canonical Representation and the definitions of the h_j's implies that

$$d(h_j'(K_j' \cap K_j''),\ h \cdot h_{j-1}(L_j - K_j)) > 0 .$$

Therefore we may use Corollary 19.1, where K_j, K_j', h_j', $h \cdot h_{j-1}(L_j - K_j)$ and $\varepsilon_1/3^{r-j-1}$ of this proof correspond to K, K', f, X and ε, respectively, of Corollary 19.1, in order to obtain a piecewise linear homeomorphism

$$h_j'':\ K_j \to E_j^n$$

such that

$$d[h_j''(K_j''),\ h\cdot h_{j-1}(L_j - K_j)] > 0;$$

$$h_j''|K_j' = h\cdot h_{j-1}|K_j' ,$$

and

$$d(h_j', h_j'') < \varepsilon_1/3^{r-j-1} .$$

Now we may define $h_j\colon L_j \to M$ as:

$$h_j(x) = \begin{cases} h\cdot h_{j-1}(x), & x \in L_j - K_j'' \\ h_j''(x), & x \in K_j \end{cases} .$$

The reader may easily verify that h_j is a well defined homeomorphism which satisfies Conditions (i) and (ii).

This completes the induction.

We note that $d(g, h_r) < \varepsilon_1 \leq \varepsilon$.

Condition (i), Lemma 15 and the fact that L is locally a k-complex imply that $h_r(L)$ is a (weakly) locally-tame set.

Thus

$$h_r\colon L_r = L \to M$$

is the desired homeomorphism, and Theorem 3 is established.

Theorems 1 and 2 follow directly from Theorems 3, 17 and 18. Corollary 1.1 is a special case of Theorem 1.

7. The Isotopies.

The bulk of this section is devoted to the

Proof of Theorem 4: This proof is by induction on the L_j's of a canonical representation of L. We will construct a set of homeomorphisms $\{f_0, f_1, \ldots, f_r;\ f_j\colon L \to E^n\}$ and a set of pushes $\{P_1, \ldots, P_r\}$ such that

i) $f_0 = 1$, $f_r = f$ and $f_j = P_j\cdot f_{j-1}$,

and

ii) $f_j|L_j = f|L_j$.

Since $f_j = P_j\cdot P_{j-1}\cdot\ldots\cdot P_1\cdot f_0$, we see that $f_j(L)$ will be (weakly) locally-tame.

Step 1. Here $j = 1$ and $L_1 = K_1$.

We observe that Lemma 15 is applicable and therefore $f(K_1)$ and K_1 are tame. We may now use Theorem 11, when $K = K_1$ and $K^* = \emptyset$, in order to obtain a push P_1 such that

$$P_1|L_1 = f|L_1 .$$

Induction hypothesis: Suppose that we have a homeomorphism f_{j-1} of L into E^n such that

(i) $f_{j-1}(L)$ is a (weakly) locally-tame set and

(ii) $f_{j-1}|L_{j-1} = f|L_{j-1}$.

We will now construct the push P_j.

Step j: We begin by observing that the hypothesis of Lemma 15 is satisfied for $f|K_j$ and $f_{j-1}|K_j$ and therefore $f|K_j$ and $f_{j-1}|K_j$ are (strongly) tame.

Therefore there is a homeomorphism h of E^n onto itself such that $h\cdot f|K_j$ is piecewise linear.

Also, if $\varepsilon_1 = d(h\cdot f_{j-1}(K_j''), h\cdot f_{j-1}(L_j - K_j))$ then Condition (8) of a canonical representation says that $\varepsilon_1 > 0$. Furthermore, we observe (check Condition (6) of a canonical representation and (ii) above) that $h\cdot f_{j-1}|K_j'$ is piecewise linear. Finally we note that $hf_{j-1}|K_j$ is (strongly) tame. Therefore we may use Theorems 12 and 9 in order to obtain an $\varepsilon_1/2$-push P_j' of $hf_{j-1}(K_j'')$ such that

$$P_j' \, hf_{j-1}|K_j \text{ is piecewise linear}$$

and

$$P_j'|hf_{j-1}(K_j') = 1 \ .$$

Therefore

(1) $$P_j'|hf(L_j - K_j) = 1 \ .$$

Furthermore, since

(2) $$hf|K_j \text{ and } P_j' \, hf_{j-1}|K_j$$

are piecewise linear, a general position argument enables us to assume with loss of generality that

(3) $$P_j' \, hf_{j-1}(K_j - K_j') \cap hf(K_j - K_j') = \emptyset$$

We will now make a minor adjustment so that we may use Theorem 13. We cannot use Theorem 13 directly because we might move some points in $hf(L_j - K_j)$ and this would be unsatisfactory.

We begin by letting K be the decomposition space of $K_j \times I$ in which the set of non-degenerate elements is

$$\{(p \times I), \ p \in K_j'\} \ .$$

We let K' be the decomposition space of $K_j \times \{0, 1\}$ in which the set of non-degenerate elements is

$$\{(p \times \{0, 1\}), \ p \in K_j'\} \ .$$

We note that K is a complex and K' is a subcomplex of K. Let K'' be the subcomplex of K induced by the decomposition of $K_j'' \times I (\subset K_j \times I)$. Thus $K = K' \cup K''$.

Let $F: K \to E^n$ be defined by

$$F(p) = hg(p) = P_j' \, hf_{j-1}(p), \quad p \in K_j' \quad ,$$

and

$$F(p \times t) = t\, hf(p) + (1-t)P_j' \, hf_{j-1}(p), \quad p \in K-K_j' \text{ and } t \in I.$$

A glance at Equations (2) and (3) shows that $F|K'$ is a piecewise linear homeomorphism.

We may now use Corollary 19.1 where the K, K', K'', f and X of the corollary correspond to K, K', K'', F and $hf(L_j-K_j)$ respectively of this theorem.

Thus we obtain a map $g: K \to E^n$ such that

$$g|K' = F|K';$$

$$g(K) \cap hf(L_j - K_j) = \emptyset \quad ,$$

and

$$d(g(K'', hf(L_j - K_j)) > 0 \ .$$

We observe that g may represent a homotopy connecting $hf|K_j$ and $P_j' \, hf_{j-1}|K_j$ which moves no point of $hf(K_j')$.

Therefore, we may apply Theorem 13 in order to obtain a push P_j'' such that

(4)
$$P_j'' \, P_j' \, hf_{j-1}|K_j = hf|K_j$$

and

$$P_j''|E^n - U_{\varepsilon_2}(g(K-K')) = 1 \quad ,$$

where we set

$$\varepsilon_2 = d(g(K'', hf(L_j - K_j)) \ .$$

Hence

(5)
$$P_j''|Hf(L_j - K_j) = 1 \quad .$$

We now check Equations (1), (4) and (5), Condition (ii) of the induction hypothesis and Condition (4) of a canonical representation in order to see that:

$$P_j'' \, P_j' \, hf_{j-1}|L_j = hf|L_j \quad .$$

Thus

$$h^{-1}P_j'' \, P_j' \, hf_{j-1}|L_j = f|L_j \quad .$$

Therefore we define

$$P_j = h^{-1} \, P_j'' \, P_j' \, h$$

and

$$f_j = P_j \cdot f_{j-1} \ .$$

We see that P_j is the desired push since

$$f_j|L_j = f|L_j \quad .$$

Thus we have shown that we may obtain the desired set of pushes $\{P_j, \ j = 1, \ldots, r\}$ and therefore the push P of the theorem is the composition $P_r \cdot P_{r-1} \cdot \ldots \cdot P_1$.

Theorem 4 is established.

Corollary 4.1 is a special case of Theorem 4.

REFERENCES

[1] R. H. Bing and J. M. Kister,"Taming complexes in hyperplanes," Duke Math. J., vol. 31 (1964) pp. 491-511.

[2] M. Brown, "A proof of the generalized Schoenflies Theorem," Bull. Amer. Math. Soc., vol. 66 (1960) pp. 74-76.

[3] J. L. Bryant, "Taming polyhedra in the trivial range," this book.

[4] J. Dancis, "Some nice embeddings in the trivial range," preceding paper in this book.

[5] H. Gluck, "Embeddings in the trivial range," Bull. Amer. Math. Soc., vol. 69 (1963) pp. 824-831.

[6] H. Gluck, "Embeddings in the trivial range (complete)," Annals of Math., vol. 81 (1965) pp. 195-210.

[7] C. A. Greathouse, "Locally flat, locally tame and tame embeddings," Bull. Amer. Math. Soc., vol. 69 (1963) pp. 820-823.

[8] V. K. A. M. Gugenheim, "Piecewise linear isotopy and embeddings of elements and spheres I, II," Proc. London Math. Soc., (3), vol. 3 (1953) pp. 29-53, 129-152.

[9] T. Homma, "On the imbeddings of polyhedra in manifolds," Yokohama Math. J., vol. 10 (1962) pp. 5-10.

ON ASPHERICAL EMBEDDINGS OF 2-SPHERES IN THE 4-SPHERE

by

Charles H. Giffen*

In 1956-7, C. D. Papakyriakopoulos proved the Loop Theorem [1], Dehn's Lemma and the Sphere Theorem [2], which have been some of the most powerful tools of 3-dimensional geometric topology. Unfortunately, such nice tools are not always available in higher dimensions—especially in dimension four. One might expect to find interesting, if not unpredictable, solutions to some 4-dimensional analogues of 3-dimensional problems whose answers involve these tools. For convenience, everything is from the semilinear point of view.

Consider the following two easy consequences of Dehn's Lemma and the Sphere Theorem:

THEOREM A (Dehn Theorem). A 1-sphere in the 3-sphere bounds a disk if and only if the fundamental group of its complement is infinite cyclic.

THEOREM B (Asphericity of knots). The complement of a 1-sphere in the 3-sphere is aspherical (i.e., its higher homotopy groups vanish).

The analogous conjectures one arrives at read as follows:

CONJECTURE A. A 2-sphere in the 4-sphere bounds a 3-cell if and only if the fundamental group of its complement is infinite cyclic.

CONJECTURE B. The complement of a 2-sphere in the 4-sphere is aspherical.

Figure 1 provides a non-locally flat counterexample to Conjecture A, so some restriction is necessary:

CONJECTURE A'. A locally flat 2-sphere in the 4-sphere bounds a 3-cell if and only if the fundamental group of its complement is infinite cyclic.

---

* Work on this paper was supported by the National Science Foundation under grant GP-3857.

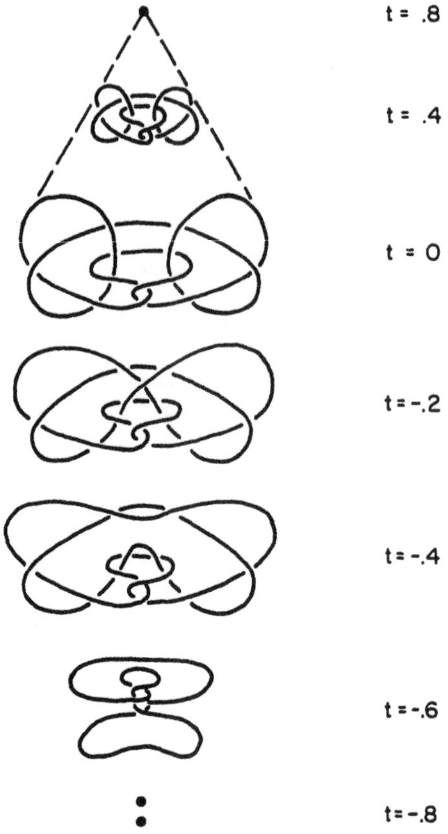

Fig. 1

In this form, the problem is unsolved and appears very difficult. A somewhat weaker conjecture than either A or A' is the following:

CONJECTURE C. The complement of a 2-sphere in the 4-sphere has the homotopy type of a circle if and only if it has infinite cyclic fundamental group.

Actually, Conjecture C is a true theorem, and an inelegant proof using combinatorial homotopy theory can be given (it is also a consequence of the principal theorem of this article). As a comparison with higher dimensions, C.T.C. Wall has shown how to tie a smooth n-sphere in the (n+2)-sphere, $n \geq 3$ whose complement has infinite cyclic fundamental group and second homotopy group isomorphic to the additive dyadic rationals!

By a result of D. B. A. Epstein [3] which shows that the second homotopy group of the complement of a spun knot is free Abelian of infinite rank if the knot is non-trivial, Conjecture B is false. Examination of the situation for spun and other fibered 2-spheres (i.e., the complementary space is fibered over a circle) leads one to the following guess:

CONJECTURE B'. The complement of a locally flat 2-sphere in the 4-sphere is aspherical if and only if it has infinite cyclic fundamental group.

By the principal theorem of this article, this conjecture is also true. One discovers that non-asphericity appears in the second homotopy group, which is generated (under the action of the fundamental group) by an element near the knotted 2-sphere. For non-locally flat 2-spheres in the 4-sphere, a non-trivial element of the second homotopy group cannot lie near the knotted 2-sphere because the boundary of a regular neighborhood is aspherical. One can eventually piece together the following conjecture:

CONJECTURE D. The complement of a 2-sphere in the 4-sphere is aspherical if and only if the inclusion induced homomorphism of fundamental groups from the boundary of a regular neighborhood of the 2-sphere to the complement of the 2-sphere is an epimorphism.

This conjecture is correct and is substantially the same as the principal theorem described in this paper. Of course Conjecture D implies Conjectures B' and C. The remainder of this article is devoted to describing the principal theorem and some of the ideas of its proof.

Let K be a 2-sphere in the 4-sphere S, with N a regular neighborhood of K, $Q = S - \text{Int } N$, $B = \text{Bd } N = \text{Bd } Q$, $G = \pi_1(B)$, $H = \pi_1(Q) = \pi_1(S-K)$, $i_*\colon G \to H$ the inclusion induced homomorphism, $J = H/\langle i_*[G, G] \rangle$ (where $\langle \ldots \rangle$ denotes smallest normal subgroup containing ...).

THEOREM. The second homotopy group $\pi_2(S-K) = \pi_2(Q)$ is the free Abelian group generated by symbols a_c, where $1 \neq c \in [J, J]$: the action of H on $\pi_2(Q)$ is that induced by the basis permutations $p_h\colon a_c \to a_{c(h)}$ where $c(h) = \pi_*(h)c\pi_*(h^{-1})$ for $h \in H$, and $\pi_*\colon H \to J$ is the natural projection; $i_*[G, G]$ is a free factor of $[H, H]$.

This is the principal theorem, and it clearly implies the conjectures claimed. The "method" of proof is to split Q into two very nice pieces and then build up the structure of $\pi_2(Q)$ from that of the two pieces and the way they fit together. More precisely, one decomposes (S, K) as the sum of two $(4, 2)$-disk pairs identified along their boundary $(3, 1)$-sphere pair in a special way:

$$(S, K) = (S_+, K_+) \cup_{(S_0, K_0)} (S_-, K_-) .$$

Then $\mathcal{Q}$ will be decomposed as the sum of the complements of open regular neighborhoods of K_+ and K_- in S_+ and S_-, respectively, identified along the complement of an open regular neighborhood of K_0 in S_0:

$$Q = \mathcal{Q}_+ \cup_{\mathcal{Q}_0} Q_- \quad ,$$

with the obvious correspondence. The particular decomposition of (S, K) will be the decomposition as the sum of two tents, identified along their boundary [4], which we briefly describe below. The meliority of this choice lies in the fact that not only is $\pi_2(\mathcal{Q}) = 0$ (by the asphericity of knots), but also $\pi_2(\mathcal{Q}_+) = 0$ and $\pi_2(Q_+, Q_0) = \mathrm{Ker}[\pi_1(\mathcal{Q}_0) \to \pi_1(Q_+)]$. Moreover, $\pi_1(B_\pm) \to \pi_1(\mathcal{Q}_\pm)$ is a monomorphism, where $B_\pm = B \cap Q_\pm$. These results follow with the help of the theory of cross-sections of a singular surface in a 4-manifold [4]. Regarding $\pi_2(Q_+, Q_0)$ as subgroups of $\pi_1(Q_0)$, each element of $\pi_2(\mathcal{Q}_+ . \mathcal{Q}_0) \cap \pi_2(\mathcal{Q}_-, \mathcal{Q}_0)$ determines precisely one element of $\pi_2(Q)$. A little more work will show that these elements generate $\pi_2(\mathcal{Q})$ and give it the required structure.

The tent decomposition is obtained as follows. A tent is a regular (4, 2)-disk pair (S_*, K_*) such that there is a general position similinear map

$$h\colon\ S_* \to [0, 1],$$

called a height function satisfying:

(0) $S_t = h^{-1}(t)$ is a point for $t = 1$ and a 3-sphere for $t < 1$;

(1) $K_* \subset h^{-1}[0, 1)$;

(2) Each non-locally flat point of K_* is an elliptic critical point, and each hyperbolic critical point is of multiplicity 2;

(3) $(h|K_*)^{-1}(\frac{1}{2}, 1)$ contains all of the elliptic points, and $(h|K_*)^{-1}(0, \frac{1}{2})$ contains all the hyperbolic points.

A theorem of [4] ensures that for any 2-sphere K in the 4-sphere S, there is a general position semilinear map

$$h\colon\ S \to [-1, 1]$$

so that $(S_+ . K_+) = (h^{-1}[0, 1], S_+ \cap K)$ and $(S_-, K) = (h^{-1}[-1, 0], S_- \cap K)$ are tents with respect to the height functions $(h|S_+)$ and $-(h|S_-)$, respectively. This gives the desired decomposition of (S, K).

It is not difficult to find a map

$$h\colon\ S \to [-1, 1]$$

so that $(h|S_+)$ and $-(h|S_-)$ satisfy (0)-(3); the trick is to make $K_0 = K \cap S_0$ connected. This is done by a process of transferring one hyperbolic point past another. This is illustrated via the contour diagram in Figure 2, and the analogue in dimension three (which cannot always be done!)

is illustrated in Figure 3. By transferring hyperbolic points suitably K_0 may be made a simple closed curve and the tent decomposition achieved.

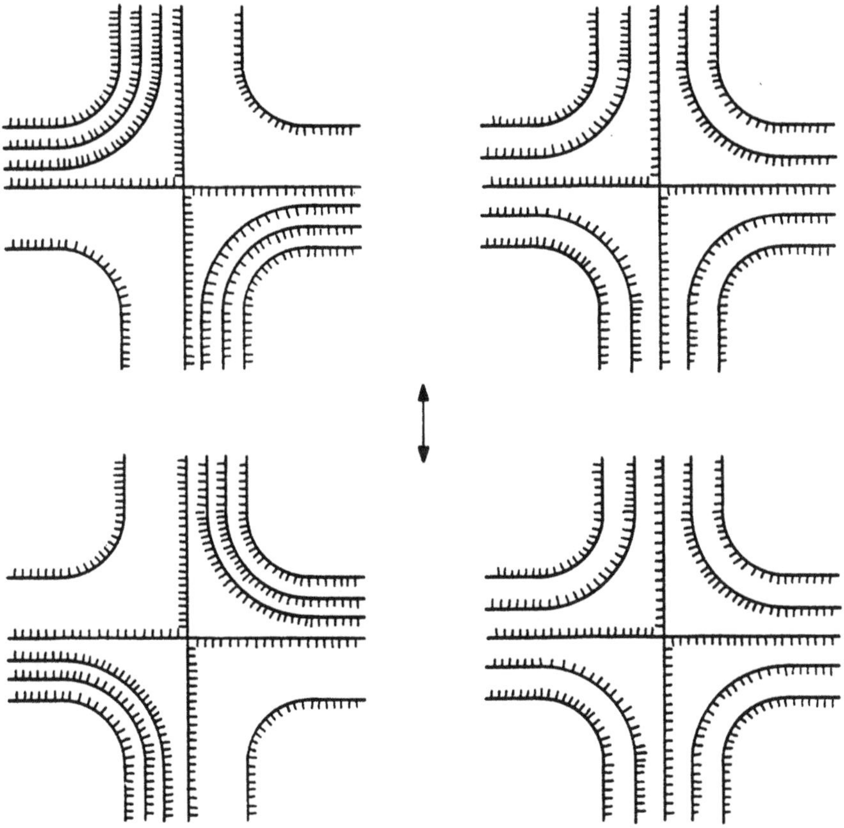

Fig. 2

If $f\colon S^2 \to \mathcal{Q}_+$ is a general position semilinear map with respect to $hf\colon S^2 \to [0, 1]$, then the same game may be played with the critical points of hf, altering f via a homotopy, so that, if elliptic point images of $(h|K_+)$ lie in $(5/6, 1)$ and hyperbolic point images of $(h|K_+)$ lie in $(1/3, 1/2)$, say, then elliptic point images of hf lie in $(0, 1/6) \cup (2/3, 5/6)$ and hyperbolic point images of hf lie in $(1/6, 1/3) \cup (1/2, 2/3)$, <u>and also</u> $(hf)^{-1}(t)$ <u>is a simple closed curve for</u> $1/3 \leq t \leq 1/2$. Once f is in this form, it is fairly easy to deform it inside $\mathcal{Q}_+$ until $hf(S^2) \subset [0, 1/3]$ where it is obviously shrinkable (since $h^{-1}[0, 1/3] \cap \mathcal{Q}_+$ is homeomorphic to $Q_0 \times [0, 1/3]$). The essential trick is that of <u>transferring a hyperbolic point of</u> hf <u>past a hyperbolic point of</u> $(h|K_+)$.

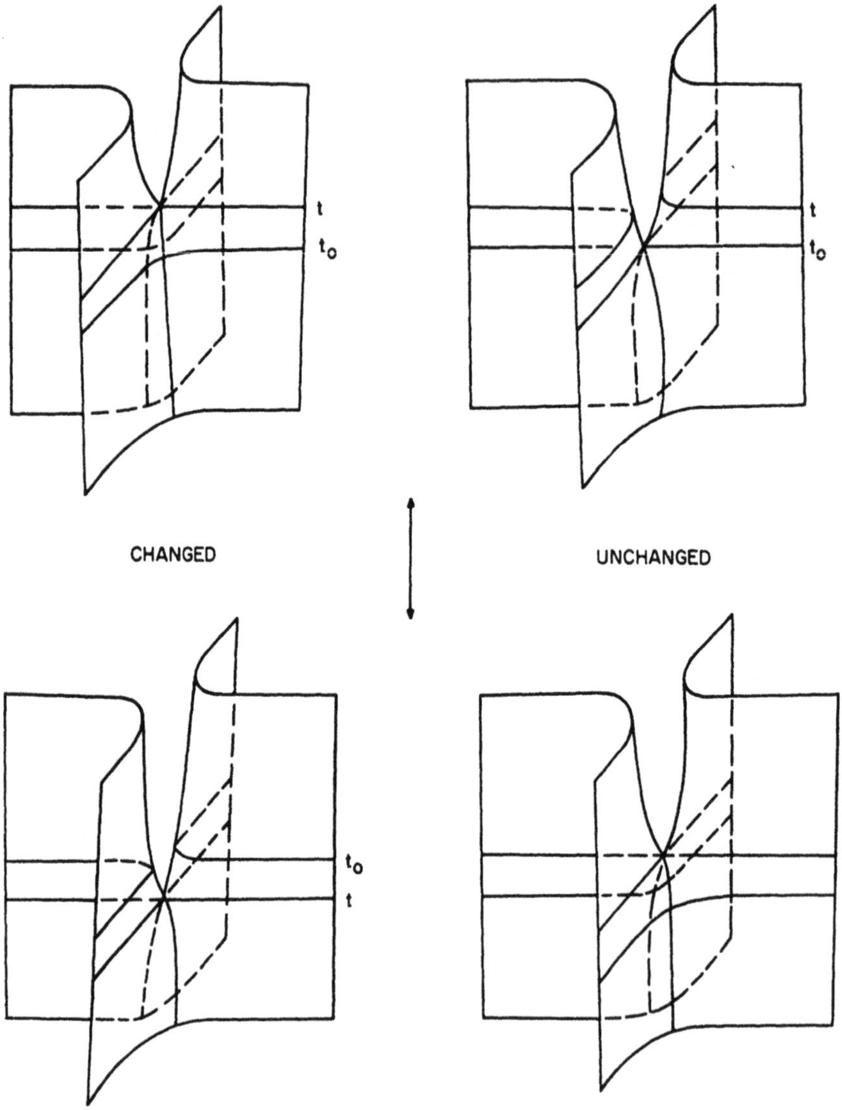

Fig. 3

The analogous construction is performed for a map $f\colon S^2 \to Q_-$. Most of the rest of the proof of the theorem uses straightforward geometric topology or careful applications of the techniques of transferring hyperbolic points.

Note: A careful study of tents (S_*, K_*) shows that $Q_* = S_* - \text{Int } N_*$, where N_* is a regular neighborhood of K_*, may be collapsed in the second derived to a 2-complex. Also, using a tent decomposition of a locally flat 2-sphere K in the 4-sphere S, one sees that K has a trivial tubular neighborhood.

REFERENCES

[1] C. D. Papakyriakopoulos, "On solid tori," Proc. London Math. Soc., (3) vol. 7 (1957) pp. 281-299.

[2] ______, "On Dehn's lemma and the asphericity of knots," Ann. of Math., vol. 66 (1957) pp. 1-26.

[3] D. B. A. Epstein, "Linking spheres," Proc. Cambridge Phil. Soc., vol. 56 (1960) pp. 215-219.

[4] C. H. Giffen, "Cross-sections of surfaces in 4-manifold," (to appear).

GEOMETRIC CHARACTERIZATION OF DIFFERENTIABLE MANIFOLDS IN EUCLIDEAN SPACE

by

Herman Gluck*

1. Introduction

In this talk I am going to describe a geometric characterization of differentiable submanifolds of class C^1 in Euclidean space. That such manifolds should yield to geometric characterization is suggested by viewing the graphs of some simple functions from R^1 to R^1. If the function is not too complicated, then one can tell visually whether it is continuous at a point, whether it is differentiable there... even whether the derivative is continuous. Beyond that... existence of second derivatives, class C^2, and so forth... visual examination loses its power. The idea behind this talk is simply that what can be observed visually should be able to be characterized geometrically.

After some preliminary definitions I will begin by discussing the problem in its simplest form... the case of one-dimensional manifolds in R^n. The results here will be simple and straightforward. The final characterization of one-dimensional C^1 manifolds in R^n, though easy to prove, will suggest (and strongly resemble) the kind of characterization we should look for in the general case. After that we will hunt for, find and prove the characterization theorem for two-dimensional C^1 manifolds in R^n. The two-dimensional case enjoys most of the essential complications of the higher dimensional analogue, but escapes some nasty technical difficulties. I will not derive the higher dimensional characterization at this time.

2. Preliminary definitions

R^n will denote n-dimensional Euclidean space.

A connected subset M of R^n is said to be a k-dimensional C^0

* The author holds an Alfred P. Sloan Research Fellowship.

manifold in R^n if for each $x \in M$ there is a neighborhood U of x in M, an open set $\mathcal{O}$ in R^k and a homeomorphism $F: \mathcal{O} \to U$.

If, in addition, F can always be chosen to be a C^1 immersion into R^n (i.e., so that F is a map of class C^1 from $\mathcal{O} \to U \subset R^n$, and the differential of F is non-singular at each point of $\mathcal{O}$), then M is said to be a k-dimensional C^1 manifold in R^n.

$G_{n,k}$ will denote the Grassman manifold of k-dimensional linear subspaces (i.e., k-planes through the origin) in R^n. It is a factor space of the orthogonal group O_n, and thereby inherits its topology. Two k-planes through the origin in R^n are close in this topology if they make a small angle with one another, i.e., if every vector in each k-plane makes a small angle with its projection on the other k-plane. Note that $G_{n,1}$ is just real projective n-1 space, P^{n-1}.

If P is a k-plane in R^n, not necessarily through the origin, then P_o will denote the k-plane through the origin which is parallel to P.

Let M be a subset of R^n and x_o a point of M. Let P be a k-plane in R^n through x_o. P is said to be a tangent k-plane to M at x_o if

$$\lim_{\substack{x \in M - x_o \\ x \to x_o}} \frac{d(x, P)}{d(x, x_o)} = 0$$

where d is the ordinary Euclidean metric in R^n. Equivalently, P is a tangent k-plane to M at x_o if for every $\varepsilon > 0$ there is a $\delta > 0$ such that if $x \in M$ and $0 < d(x, x_o) < \delta$, then the vector $x - x_o$ makes an angle less than ε with the k-plane P. If M has a tangent k-1-plane P at x_o, then any k-plane containing P is a tangent k-plane to M at x_o. On the other hand, if M has no tangent k-1-plane at x_o, then it has at most one tangent k-plane at x_o.

3. One-dimensional manifolds in R^n

The most obvious property of a one-dimensional C^1 manifold in R^n, and a reasonable first choice for a characterization, is the possession by such a manifold of a continuously turning tangent line. This is really a double barrelled proposition. If M is a one-dimensional C^1 manifold in R^n, then

(1) M has a tangent line $L(x)$ at each point $x \in M$, and

(2) the map from $M \to G_{n,1}$ which sends $x \to L_o(x)$ is continuous.

This property does not, however, characterize one-dimensional C^1 manifolds. For example, the curve $y^3 = x^2$ in R^2 is a one-dimensional C^o manifold in R^2 which has a continuously turning tangent line, but is not a C^1

manifold. The difficulty, clearly enough, is the cusp at the origin.

There is an additional property of one-dimensional C^1 manifolds in R^n which has the effect of outlawing cusps. If M is a one-dimensional C^1 manifold in R^n and x is a point of M, let π_x denote the orthogonal projection of R^n onto the tangent line $L(x)$ to M at x. Then

(3) π_x is one-one (and therefore a homeomorphism into) on a sufficiently small neighborhood of x in M.

The proof of this is a simple application of the mean value theorem. Note that the tangent line to the curve $y^3 = x^2$ at the origin is just the y-axis, and that the orthogonal projection of R^2 onto the y-axis is 2-to-1 on arbitrarily small neighborhoods of the origin on the curve. It is in this way that property (3) outlaws cusps.

It turns out that properties (1), (2) and (3) do characterize one-dimensional C^1 manifolds in R^n. Specifically, we have

THEOREM 3.1. Let M be a one-dimensional C^0 manifold in R^n. Then M is a C^1 manifold in R^n if and only if

(1) M has a tangent line $L(x)$ at each point $x \in X$;

(2) the map $M \to G_{n,1}$ sending $x \to L_0(x)$ is continuous;

(3) for each $x \in M$, the orthogonal projection π_x of R^n onto $L(x)$ is one-one (and therefore a homeomorphism into) on some neighborhood U of x in M.

Pick a point $x \in M$ and assume for simplicity that x is the origin of R^n and that the tangent line $L(x)$ coincides with $R^1 \subset R^n$. If M really is a C^1 manifold, then the local parametrization $(\pi_x/U)^{-1}$ will turn out to be a C^1-immersion. The idea of the proof is simply to use properties (1), (2) and (3) to prove directly that $(\pi_x/U)^{-1}$ is a C^1-immersion, from which it will follow (since x was an arbitrary point in M) that M is a C^1 manifold. The details are dull and I will omit them.

While the above theorem _is_ a characterization of one-dimensional C^1 manifolds in Euclidean space, it probably leaves the reader a bit cold. This is undoubtedly because the use of a hypothesis about the existence of a tangent line (as geometrical as this notion may be) in an attempt to characterize a C^1 manifold gives the appearance of begging the question. The above theorem is, in any case, not the sort of characterization I have in mind, but it (and eventually its higher dimensional analogue) will turn out to be a very useful tool.

In order to hit on the characterization I do have in mind, let us consider yet another property of a one-dimensional C^1 manifold M in R^n.

Let Δ denote the diagonal of $M \times M$, i.e., the set of points $\{(x, x): x \in M\}$, and consider the secant map

$$\Sigma: \; M \times M - \Delta \to G_{n,1}$$

which associates with the pair (x, y) the line in R^n through the origin and parallel to the vector $x-y$. The secant map Σ is certainly continuous... in fact, it would be continuous for any subset M of R^n. But because M is a one-dimensional C^1 manifold, we have

(4) the secant map $\Sigma: M \times M - \Delta \to G_{n,1}$ has a continuous extension over all of $M \times M$ by defining $\Sigma(x, x)$ to be $L_0(x)$, the line through the origin parallel to the tangent line to M at x.

The proof of this is, as usual, an application of the mean value theorem.

Property (4) provides precisely the kind of characterization we are looking for.

THEOREM 3.2. Let M be a one-dimensional C^0 manifold in R^n. Then M is a C^1 manifold in R^n if and only if the secant map $\Sigma: M \times M - \Delta \to G_{n,1}$ admits a continuous extension over all of $M \times M$.

We have already observed the necessity of the condition. To prove sufficiency, let M be a one-dimensional C^0 manifold in R^n whose secant map admits a continuous extension over $M \times M$. We will show that M is a C^1 manifold by showing that M satisfies conditions (1), (2) and (3), and then invoking Theorem 3.1.

For each $x \in M$, let $L(x)$ be the line in R^n passing through x and parallel to the line $\Sigma(x, x)$. The continuity of Σ on $M \times M$ implies that for every $\varepsilon > 0$ there exists a $\delta > 0$ such that if y is a point of M satisfying $0 < d(x, y) < \delta$, then the line $\Sigma(x, y)$ makes an angle $< \varepsilon$ with $\Sigma(x, x)$, and hence with $L(x)$. Thus $L(x)$ is a tangent line to M at x, verifying condition (1).

Condition (2), the fact that $L_0(x) = \Sigma(x, x)$ varies continuously with x, follows directly from the continuity of Σ on $M \times M$.

Given $x \in M$, use the continuity of Σ to get a neighborhood U of x in M such that $\Sigma(y, z)$ makes an angle $< \pi/2$ with $\Sigma(x, x)$ for all pairs $(y, z) \in U \times U$. Then the orthogonal projection π_x of R^n onto $L(x)$ is one-one on U, verifying condition (3).

Theorem 3.1 is now applicable, and the proof is completed.

In the following sections I shall try to find and prove a characterization theorem, in the spirit and form of the above result, for higher dimensional manifolds in Euclidean space.

To close this section, the following theorem characterizes those subsets of R^n which are one-dimensional C^1 manifolds, without the preliminary assumption (which we have relied on up to this point) that we are at least dealing with a C^0 manifold. No generalization of this result will be attempted in the following sections.

THEOREM 3.3. A subset $M \subset R^n$ is a one-dimensional C^1 manifold in R^n if and only if M is connected, has no local non-cut points, and the secant map Σ: $M \times M - \Delta \to G_{n,1}$ admits a continuous extension over all of $M \times M$.

The conditions are of course necessary. To prove sufficiency, we need only show from the conditions that M is a one-dimensional C^0 manifold and then invoke Theorem 3.2.

Let x be a fixed but arbitrary point of M and let $L(x)$ be the line through x parallel to $\Sigma(x, x)$. Let U be a neighborhood of x in M, having the form of an intersection between M and a cylindrical neighborhood of x in R^n with axis $L(x)$, and chosen so that $(y, z) \in U \times U$ will imply that $\Sigma(y, z)$, and hence the vector $y-z$, makes an angle $< \pi/4$ with $L(x)$. This will in particular force the orthogonal projection π_x of R^n onto $L(x)$ to be one-one on U. Let I denote the image $\pi_x(U)$.

4. Two-dimensional manifolds in R^n

If we proceed along the lines set in the last section, then the most obvious property of a two-dimensional C^1 manifold in R^n is the possession of a continuously turning tangent plane. Specifically, if M is a two-dimensional C^1 manifold in R^n, then

(1) M has a tangent plane $P(x)$ at each point $x \in M$, and

(2) the map from $M \to G_{n,2}$ which sends $x \to P_0(x)$ is continuous.

As we expect, this property does not characterize two-dimensional C^1 manifolds in R^n. For example, the subset of R^3

$$\{(x, y, z): \; y^3 = x^2\}$$

is a two-dimensional C^0 manifold in R^3 with a continuously turning tangent plane, but it is not a C^1 manifold. This example is nothing but the corresponding one-dimensional example, "multiplied by a line."

If M is a two-dimensional C^1 manifold in R^n and x is a point of M, let π_x denote the orthogonal projection of R^n onto the tangent plane $P(x)$ to M at x. Then the ever-useful mean value theorem yields

(3) π_x is one-one (and therefore a homeomorphism into) on a sufficiently small neighborhood of x in M.

As in the one-dimensional case, this property outlaws the cusp-like behavior of the example given above.

Just as in the one-dimensional case, properties (1), (2) and (3) characterize two-dimensional C^1 manifolds in R^n.

THEOREM 4.1. Let M be a two-dimensional C^0 manifold in R^n. Then M is a C^1 manifold in R^n if and only if

(1) M has a tangent plane $P(x)$ at each point $x \in M$;
(2) the map $M \to G_{n,2}$ sending $x \to P_0(x)$ is continuous;
(3) for each $x \in M$, the orthogonal projection π_x of R^n onto $P(x)$ is one-one (and therefore a homeomorphism into) on some neighborhood U of x in M.

As in the one-dimensional case, the idea of the proof is to use the above properties to show directly that $(\pi_x/U)^{-1}$ is a C^1-immersion. The proof is straightforward, but somewhat more involved than its one-dimensional prototype. I will omit the details. This theorem will prove to be a useful tool.

Remark: With the obvious modifications, the above theorem is true (and just as easily proved) for k-dimensional manifolds in R^n.

5. The shape of a triangle

We come now to the heart of the story. Let M be a two-dimensional C^1 manifold in R^n. By analogy with the one-dimensional case, we should now consider triples (u, v, w) of distinct points of M, pass a plane $P(u, v, w)$ through these three points, and expect that as u, v and $w \to x \in M$, $P_0(u, v, w)$ will approach $P_0(x)$. But it just isn't so! Simply look at three distinct points u, v and w on the equator of a two-sphere, and let them all approach a single point x on this equator. The plane $P(u, v, w)$ always contains the equator, and is hence always orthogonal to $P(x)$, the tangent plane to the two-sphere at x. So $P_0(u, v, w)$ hasn't a ghost of a chance of approaching $P_0(x)$.

It is precisely this sort of behavior which makes the study of surface area so much more difficult than the study of arc length. Schwarz's example of a sequence of polygonal approximations to a finite cylinder, in which the surface areas of the approximating polygons grow without bound, is based on just this phenomenon of small triangles inscribed in a surface making a large angle with nearby tangent planes to the surface.

What is it that goes wrong? If u, v and w are distinct points on M close to x, the mean value theorem will imply that each edge of the triangle uvw makes a small angle with the tangent plane $P(x)$ to M at x. But here is the rub: just because the three edges of a triangle make a small angle with a given plane, it does not follow that the triangle itself (i.e., the plane of the triangle) makes a small angle with the given plane.

Can the situation be controlled so that things do work out right? It is easy to calculate, for example, that if the three edges of an equilateral triangle make a small angle with a planê, then the triangle itself makes a small angle with the plane. In some sense, the closer the "shape" of a triangle is to "degeneracy" (all three vertices lying on a straight line), the larger the angle that triangle can make with a given plane, even though its three edges make small angles with the plane. In order to obtain the kind of characterization of two-dimensional C^1 manifolds in R^n that I have in mind, we will have to first prove a lemma to the effect that if a triangle is not too degenerate, then small angles between its edges and a fixed plane imply a small angle between the triangle itself and the fixed plane.

Before I give the precise statement of this lemma, let me mention that my approach to the characterization of two-dimensional C^1 manifolds in R^n is modelled after a similar approach taken by Toralballa [1] to problems in the theory of surface area. In fact, the ideas that I am now expounding are the products of several conversations with Professor Toralballa.

Now for the statement of the lemma alluded to above.

LEMMA 5.1. Let uvw be a triangle in R^n, each of whose edges makes an angle $< \varepsilon$ with some unknown plane P^*. If θ is the largest angle in the triangle, then the angle α between the plane P of the triangle and the plane P^* is bounded by the condition

$$\sin \alpha < \frac{\sin \varepsilon}{\cos(\theta/2)} .$$

This bound is best possible, in the sense that one can find a triangle in R^3 each of whose edges makes an angle $\leq \varepsilon$ with a plane P^*, while the angle α between the plane P of the triangle and P^* satisfies

$$\sin \alpha = \min\left(\frac{\sin \varepsilon}{\cos(\theta/2)} , 1\right) .$$

The proof is an exercise in geometry, and will be omitted.

We can now define a shape function σ for triangles, as follows. Let u, v and w be three distinct points in R^n. Then, even if they happen to be colinear, one can still speak of the angles of the triangle uvw. Let θ be the largest angle of this triangle. Then we define

$$\sigma(uvw) = \cos(\theta/2) .$$

The shape function σ is continuous on the space or ordered triples of distinct points in R^n, and is bounded by

$$0 \leq \sigma \leq \sqrt{3}/2 \approx .866 .$$

A triangle uvw with three distinct vertices is degenerate (the worst possible shape) if and only if $\sigma(uvw) = 0$, and is equilateral (the best possible shape) if and only if $\sigma(uvw) = \sqrt{3}/2$.

If the reader is mildly disturbed that the shape function of an equilateral triangle is only $\sqrt{3}/2$ (after all, an equilateral triangle is a perfect triangle, hence its shape function ought to have the value 1), let me remark that for our purposes, an equilateral triangle is not perfact. A triangle would be perfect if, when told that each edge made an angle $< \varepsilon$ with an unknown plane P^*, one could conclude that the plane P of the triangle also made an angle $< \varepsilon$ with P^*. But Lemma 5.1 tells us that no triangle is perfect. In fact, the best we can say is that if each angle of the non-degenerate triangle uvw makes an angle $< \varepsilon$ with the plane P^*, then the plane P of the triangle makes an angle α with P^* subject to the bound

$$\sin \alpha < \frac{\sin \varepsilon}{\sigma(uvw)} .$$

The above definition of the shape function σ can be given the following interpretation. Consider three distinct points u, v and w in R^n and the resulting triangle uvw. For simplicity, assume $n = 2$ (the points u, v and w will always lie on a two-dimensional subspace of R^n anyway). Let L, M and N be lines through the origin in R^2 parallel to the edges of the triangle uvw. Then L, M and N are elements of the projective line P^1. The triangle uvw has "good" shape if the directions of its edges are "well-distributed," i.e., if L, M and N are "well-distributed" over P^1. The most straightforward measure of the distribution of L, M and N over P^1 is the maximum distance from any point of P^1 to the set $\{L, M, N\}$. If we use angular measurement on P^1, then this maximum distance is precisely $\theta/2$, where θ is the largest angle of the triangle uvw. So the number $\theta/2$ measures the shape of the triangle uvw. The shape function as we have actually defined it is $\cos(\theta/2)$ rather than $\theta/2$, because this fits better with Lemma 5.1, and because it assigns the number 0 as the shape of a degenerate triangle.

Let $(R^n)^3 = R^n \times R^n \times R^n$ denote the set of ordered triples of points in R^n. For each δ, $0 \leq \delta \leq \sqrt{3}/2$, let $(R^n)^3_\delta$ denote the set of ordered triples (u, v, w) of distinct points in R^n satisfying $\sigma(uvw) \geq \delta$.

Remarks: (1) $(R^n)^3_0$ is the set of all ordered triples of distinct points in R^n. $(R^n)^3_{\sqrt{3/2}}$ is the set of all ordered triples of points in R^n which form the vertices of an equilateral triangle.

(2) If $\delta \leq \delta'$, then $(R^n)^3_\delta \supset (R^n)^3_{\delta'}$.

(3) $(R^n)^3_\delta$ is closed in $(R^n)^3_0$ because σ is a continuous function on $(R^n)^3_0$, but it is not closed in $(R^n)^3$.

(4) For any value of δ, $0 \leq \delta \leq \sqrt{3/2}$, the closure of $(R^n)^3_\delta$ in $(R^n)^3$ will contain the diagonal $\Delta = \{(x, x, x): x \in R^n\}$. The closure of $(R^n)^3_\delta$ will be precisely $(R^n)^3_\delta \cup \Delta$ when $\delta > \sqrt{2/2}$. But when $0 \leq \delta \leq \sqrt{2/2}$, the closure of $(R^n)^3_\delta$ will be $(R^n)^3_\delta \cup \Delta^*$, where $\Delta^* = \{(u, v, w) \in (R^n)^3: u = v \text{ or } v = w \text{ or } w = u\}$.

(5) Fix $\delta > 0$. If $(u, v, w) \in (R^n)^3_\delta$ and each edge of the triangle uvw makes an angle $< \varepsilon$ with a plane P^*, then the plane P of the triangle makes an angle α with P^* subject to the bound

$$\sin \alpha < \frac{\sin \varepsilon}{\delta} .$$

Thus as $\varepsilon \to 0$, so does $\alpha \to 0$.

6. The distribution of triangles of good shape on a two-dimensional C^0 manifold

LEMMA 6.1. Let M be a two-dimensional C^0 manifold in R^n, x_0 a point of M and W a neighborhood of x_0 in M. Then there is a neighborhood U of x_0 in M such that if u and v are distinct points of U, there is a thrid point $w \in W$ such that the largest angle of the triangle uvw is $\leq 90°$, i.e., such that $\sigma(uvw) \geq \sqrt{2/2}$.

Let M, x_0 and W be given as above. Making W smaller if necessary, we may assume that W is connected. Let U be any neighborhood of x_0 in M subject to the conditions:

(i) $U \subset W$

(ii) $\operatorname{diam} U < \operatorname{diam} W$.

Now let u and v be any two distinct points selected from U. We will construct the set $A \subset R^n$ of all points w, distinct from u and v, which together with u and v form a triangle uvw whose largest angle is $\leq 90°$. To prove the theorem, we will simply show that the original neighborhood W must intersect the set A.

Draw through the point u the $n-1$ dimensional hyperplane orthogonal to the line uv, as shown below. The third vertex w must lie on or

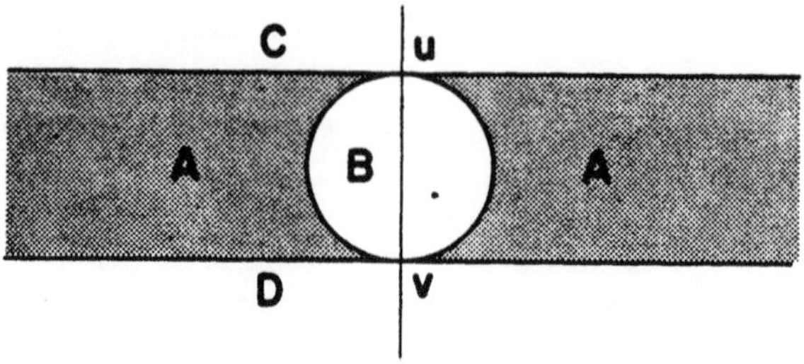

"below" this hyperplane (see figure) if the triangle uvw is to have an angle $\leq 90°$ at u. Similarly, w must lie on or "above" the n-1 dimensional hyperplane through v and orthogonal to the line uv, if the triangle is to have an angle $\leq 90°$ at v. Finally, w must lie on or outside an n-1 sphere of radius $(d(u, v))/2$ with center at $(u+v)/2$ if the triangle uvw is to have an angle $\leq 90°$ at w. The combination of these three restrictions yields the set A, as shown in the above figure. The points u and v are the only limit points of A not actually contained in A.

We must now show that the neighborhood W intersects the set A. Assume, to the contrary, that $W \cap A = \emptyset$. Then (see figure) $W \subset B \cup C \cup d \cup \{u, v\}$. W cannot be entirely contained in $B \cup \{u, v\}$, because $\operatorname{diam} W > \operatorname{diam} U \geq d(u, v)$, while the diameter of $B \cup \{u, v\}$ is just $d(u, v)$. W cannot be entirely contained in $C \cup D \cup \{u, v\}$, because u cannot be connected to v in $C \cup D \cup \{u, v\}$, whereas it can in W. So, under the assumption that $W \cap A = \emptyset$, we have W contained entirely in $B \cup C \cup D \cup \{u, v\}$, partially in B, partially in $C \cup D$ and (of course) partially in $\{u, v\}$. But then removal of either u or v from W disconnects W, whereas this is patently impossible in a two-dimensional manifold. So W must meet A, and therefore there is a point $w \in W$ such that the largest angle of the triangle uvw is $\leq 90°$, proving the lemma.

<u>Remark</u>: The above lemma is best possible, i.e., if we substitute for 90° any smaller angle, the lemma becomes false.

Let $(M)^3 = M \times M \times M$ denote the set of ordered triples of points in the two-manifold M. Let $(M)^3_\delta = (M)^3 \cap (R^n)^3_\delta$ denote the set of ordered triples (u, v, w) of distinct points of M such that $\sigma(uvw) \geq \delta$. Let Δ denote the diagonal of $(M)^3$.

COROLLARY TO LEMMA 6.1. Let M be a two-dimensional C^0 manifold in R^n and δ a real number satisfying $0 \leq \delta \leq \sqrt{2}/2$. Then the closure of $(M)^3_\delta$ contains the diagonal Δ of $(M)^3$.

7. Characterization of two-dimensional C^1 manifolds in R^n

Let M be a two-dimensional C^1 manifold in R^n, and consider the secant map

$$\Sigma: \quad (M)^3_o \to G_{n,2}$$

which associates with the triple (u, v, w) of distinct points in M the plane $P_o(u, v, w)$ through the origin in R^n and parallel to the plane $P(u, v, w)$ through the three points u, v and w. The map Σ is certainly continuous but, as we saw in section 5, there is no hope of extending it to a continuous map over $(M)^3_o \cup \Delta$. But we can now get around this predicament! Pick any real number δ subject to the condition $0 < \delta < \sqrt{3}/2$, and restrict the secant map to $(M)^3_\delta$,

$$\Sigma: \quad (M)^3_\delta \to G_{n,2} \quad .$$

Now define $\Sigma(x, x, x)$ to be $P_o(x)$, the plane through the origin parallel to the tangent plane to M at x. As $(u, v, w) \in (M)^3_\delta$ approaches (x, x, x), the mean value theorem tells us that the edges of the triangle uvw make very small angles with $P_o(x)$. But if these three edges each make an angle $< \varepsilon$ with $P_o(x)$, then the plane $P_o(u, v, w)$ makes an angle α with $P_o(x)$ subject to the bound

$$\sin \alpha < \frac{\sin \varepsilon}{\delta} ,$$

as was observed in Remark (5) on page 202. But then as $\varepsilon \to 0$, so does $\alpha \to 0$. Hence $P_o(u, v, w) = \Sigma(u, v, w)$ approaches $P_o(x) = \Sigma(x, x, x)$. Thus the extended map Σ,

$$\Sigma: \quad (M)^3_\delta \cup \Delta \to G_{n,2}$$

is continuous. So we have established

(4) for each real number δ, $0 < \delta \leq \sqrt{3}/2$, the secant map $\Sigma: \ (M)^3_\delta \to G_{n,2}$ has a continuous extension over $(M)^3_\delta \cup \Delta$ by defining $\Sigma(x, x, x) = P_o(x)$, the plane through the origin parallel to the tangent plane to M at x.

This is the generalization of item (3) of section 3.

We are ready now to give the characterization of two-dimensional C^1 manifolds in R^n.

THEOREM 7.1. Let M be a two-dimensional C^o manifold in R^n and δ a real number, $0 < \delta \leq \sqrt{2}/2$. Then M is a C^1 manifold if and only if the secant map $\Sigma: \ (M)^3_\delta \to G_{n,2}$ admits a continuous extension over $(M)^3_\delta \cup \Delta$.

We have just observed, in item (4) above, the necessity of the condition under the more liberal bounds: $0 < \delta \leq \sqrt{3}/2$.

To prove sufficiency, let M be a two-dimensional C^0 manifold in R^n whose secant map admits a continuous extension over $(M)^3_\delta \cup \Delta$, $0 < \delta \leq \sqrt{2}/2$. We will show that M is a C^1 manifold by showing that M satisfies conditions (1), (2) and (3) of Theorem 4.1, and then invoking that theorem.

For each $x \in M$, let $P(x)$ be the plane in R^n passing through x and parallel to the plane $\Sigma(x, x, x)$. To show that $P(x)$ is the tangent plane to M at x, we must show that for $y \in M - \{x\}$ sufficiently close to x, the vector $x-y$ makes a small angle with $P(x)$. To do this, use the continuity of Σ on $(M)^3_\delta \cup \Delta$ to choose a neighborhood W of x such that if u, v and w are three distinct points in W and $(u, v, w) \in (M)^3_\delta$, then $\Sigma(u, v, w)$ makes a small angle with $\Sigma(x, x, x)$. Now use Lemma 6.1 to select a smaller neighborhood U of x such that if u and v are distinct points in U, there will exist a thrid point $w \in W$ such that $\sigma(uvw) \geq \sqrt{2}/2 \geq \delta$, and hence $(u, v, w) \in (M)^3_\delta$. Finally, let $u = x$ and $v = y$. Then there is a third point $z \in W$ such that $(x, y, z) \in (M)^3_\delta$. Hence $\Sigma(x, y, z)$ Makes a small angle with $\Sigma(x, x, x)$, and hence with $P(x)$. But then the vector $x-y$, being parallel to an edge of the triangle xyz, makes at least as small an angle with $P(x)$. So $P(x)$ is indeed a tangent plane to M at x. This verifies condition (1) of Theorem 4.1.

Condition (2), the fact that $P_0(x) = \Sigma(x, x, x)$ varies continuously with x, follows directly because Σ is continuous on $(M)^3_\delta \cup \Delta$, and hence surely on Δ.

Condition (3) follows in the same fashion as did condition (1). Given $x \in M$, pick a neighborhood W of x in M so small that if u, v and w are three points in W with $(u, v, w) \in (M)^3_\delta$, then $\Sigma(u, v, w)$ makes an angle ε with $\Sigma(x, x, x)$ such that

$$\frac{\sin \varepsilon}{\delta} < 1.$$

Following the pattern of the argument for condition (1), we use Lemma 6.1 to get a smaller neighborhood U of x such that if u and v are distinct points in U, there will exist a thrid point $w \in W$ such that $(u, v, w) \in (M)^3_\delta$. Then by Remark (5) on page , the plane $\Sigma(u, v, w)$, and hence surely the vector $u-y$, makes an angle $< 90°$ with the plane $\Sigma(x, x, x)$. This implies that the orthogonal projection π_x of R^n onto the tangent plane $P(x)$ is one-one on U, verifying condition (3).

Theorem 4.1 is now applicable, and completes the proof of the theorem.

REFERENCES

[1] L. V. Toralballa, Surface Area (to appear).

HARVARD UNIVERSITY

WHITEHEAD TORSION AND h-COBORDISM

by

R. H. Szczarba[†]

Introduction

In this lecture, we will be concerned with the properties of h-cobordisms, particularly the relationship between h-cobordisms and Whitehead torsion. Most of what we say here holds equally well in either the differentiable or piecewise linear category so we will use "manifold" to mean either differentiable or piecewise linear manifold. Two manifolds will be termed <u>isomorphic</u> if they are either diffeomorphic or piecewise linearly homeomorphic and we write $M \approx M'$. Finally, all manifolds will be compact, oriented unless explicitly stated otherwise, ∂M will denote the boundary of M, and $-M$ will denote M with its opposite orientation.

The material here is divided into three sections. The first recalls the h-cobordism theorem of Smale and its generalizations, the second deals with the Whitehead group and the notion of torsion, and the last contains some properties of h-cobordisms.

1. The h-cobordism and s-corbordism theorems.

A triad of manifolds $(W; M, M')$ is an h-cobordism if ∂W is the disjoint union $M \cup (-M')$ and each of the inclusions $M \subset W$, $M' \subset W$ is a homotopy equivalence. (It will sometimes be convenient to refer to W as an h-cobordism.) We say M and M' are h-<u>cobordant</u> if there is an h-cobordism $(W; M, M')$. The fundamental result of Smale [8] is the following (see also Milnor [7]).

"h-<u>cobordism</u>" <u>Theorem</u>. Let $(W; M, M')$ be a differentiable h-cobordism with $\pi_1 W = 0$ and $\dim W \geq 6$. Then W is diffeomorphic to $M \times I$, $I = [0, 1]$. In particular, M is diffeomorphic to M'.

To demonstrate the strength of this result, let Σ be a differentiable homotopy n-sphere and $W = \Sigma - (D_1 \cup D_2)$ where D_1 and D_2 are disjoint n-discs in Σ. It is easily checked that W is an h-cobordism

† The author was partially supported by NSF-GP-4037.

so, if $n \geq 6$, W is diffeomorphic to $S^{n-1} \times I$, $S^{n-1} = \partial D_1$. Thus, $\Sigma \approx (S^{n-1} \times I) \cup D_1 \cup D_2$ which is clearly homeomorphic to S^n and we have proved the Poincaré conjecture in dimensions ≥ 6.

If the hypothesis $\pi_1 W = 0$ is omitted from the h-cobordism theorem, it no longer is true. In order to deal with the non simply connected case, it is necessary to define the Whitehead group $Wh(\pi)$ of a discrete group π and to associate with any h-cobordism $(W; M, M')$ a torsion element $\tau(W, M)$ in $Wh(\pi_1 W)$. We will call an h-cobordism $(W;M,M')$ an s-cobordism if $\tau(W, M) = 0$. The following is due to Mazur [5], Barden [1], and Stallings [9].

"s-cobordism" Theorem. Let $(W; M, M')$ be an s-cobordism with $\dim W \geq 6$. Then W is isomorphic to $M \times I$.

Remark. Since $Wh(\pi) = 0$ if $\pi = 0$, the s-cobordism theorem does generalize the h-cobordism theorem.

We close this section with a strengthened form of the s-cobordism theorem due to Stallings [9]. An application will be given in Section 3.

THEOREM. Let M be a manifold of dimension ≥ 5 and α any element of $Wh(\pi_1 M)$. Then there exists an h-cobordism $(W_\alpha; M, M_\alpha)$ unique up to isomorphism such that $\tau(W_\alpha, M) = j_\# \alpha$ where $j_\#: \pi_1 M \to \pi_1 W_\alpha$ is induced by $j: M \subset W_\alpha$.

2. The Whitehead group and the notion of torsion

In this section, we define the Whitehead group of a ring and of a discrete group and outline the procedure for defining the torsion of a homotopy equivalence and of an h-cobordism. Two general references for the material in this section are Whitehead [10] and the excellent set of notes of Milnor [6].

Let R be an associative ring with unit having the property that any two bases for a finitely generated from R-module have the same number of elements. Let $GL(n, R)$ be the group of invertible $n \times n$ R-matrices and $GL(R) = \bigcup_{n \geq 1} GL(n, R)$. We do not distinguish between an element of $GL(n, R)$ and the element of $GL(R)$ determined by it.

Let M be the subgroup of $GL(R)$ generated by those matrices which are the identity except for a single non-zero off diagonal entry together with the single 1×1 matrix (-1). The reduced Whitehead group of R, $\overline{K}_1(R)$, is defined to be the quotient $GL(R)/M$. According to Whitehead [10], $\overline{K}_1(R)$ is an Abelian group.

Let π be a discrete group and $R = Z[\pi]$ its integral group ring. Let M' be the subgroup of $GL(R)$ generated by M and the 1×1 matrices (g), $g \in \pi$. The Whitehead group of π, $Wh(\pi)$, is defined to be the quotient $GL(R)/M'$. Clearly $Wh(\pi)$ is a homomorphic image of $\overline{K}_1(Z[\pi])$.

The following facts are known about $Wh(\pi)$.

(i) $Wh(\pi) = 0$ if $\pi \approx Z_2, Z_3, Z_4$, or Z (see Higman [3]) and $Wh(Z_5) \approx Z$. Here $Z_m = Z/mZ$.

(ii) If π if finite Abelian, $Wh(\pi)$ is a finitely generated free Abelian group. (See Bass [2].)

We now define the torsion of an h-cobordism in four steps.

Step 1. Let C be an R-complex with the property that each C_q is a finitely generated free R-module with a prescribed R-basis and $C_q = 0$ for $q > N$, some N. Such a complex will be called a based R-complex. If C is a based, acyclic R-complex, the torsion of C, $\tau(C) \in \overline{K}_1(R)$ is uniquely determined by the following two properties.

a) If $C_q = 0$ for $q \neq n+1, n$, then $\tau(C)$ is the element of $\overline{K}_1(R)$ determined by the matrix defining the isomorphism $\partial\colon C_{n+1} \to C_n$ (with a sign $(-1)^n$).

b) If $0 \to C' \to C \to C'' \to 0$ is a short exact sequence of based acyclic R-complexes with the obvious compatibiltiy condition on bases, then $\tau(C) = \tau(C') + \tau(C'')$.

Step 2. If $\varphi\colon C \to C'$ is a chain equivalence between based R-complexes, then the algebraic mapping cone of φ, C_φ, is the based, acyclic R-complex defined by $(C_\varphi)_q = C_{q-1} \oplus C'_q$, $\partial_\varphi(C, C') = (\partial C, \varphi C - \partial' C')$. We define the torsion of φ by $\tau(\varphi) = \tau(C_\varphi)$.

Step 3. Suppose X and Y are finite simplicial complexes and $f\colon X \to Y$ a simplicial homotopy equivalence. If $\tilde{X}$ and $\tilde{Y}$ are the universal covering spaces and $\tilde{f}\colon \tilde{X} \to \tilde{Y}$ the homotopy equivalence induced by f, then $\tilde{f}_\#\colon C(\tilde{X}) \to C(\tilde{Y})$ is a chain equivalence of $Z[\pi_1 Y]$-modules. (Here $C(\tilde{Y})$ becomes a $Z[\pi_1 Y]$-module by identifying $\pi_1 Y$ with the group of covering transformations on $\tilde{Y}$ and $C(\tilde{X})$ by using the isomorphism $f_\#\colon \pi_1 X \to \pi_1 Y$.) Now if we choose for each q-simplex σ of Y a q-simplex $\tilde{\sigma}$ of $\tilde{Y}$ lying over it, we obtain a $Z[\pi_1 Y]$-base for $C(\tilde{Y})$. Doing the same for $C(\tilde{X})$, $\tilde{f}_\#$ becomes a chain equivalence of based complexes so we can define $\tau(\tilde{f}_\#) \in \overline{K}_1(Z[\pi_1 Y])$. We define the torsion of f, $\tau(f)$, to be the element of $Wh(\pi_1 Y)$ determined by $\tau(\tilde{f}_\#)$. It is easily checked that $\tau(f)$ does not depend on the choice of the simplices $\tilde{\sigma}$.

If $\tau(f) = 0$, we say that f is a simple equivalence and that X and Y have the same simple homotopy type.

Remark 1. The geometric meaning of simple homotopy type is as follows. Let σ be a simplex of X with exactly one free face. The operation of passing from X to X - interior σ is called an elementary contraction and the inverse operation an elementary expansion. Whitehead proves in [10] that X and Y have the same simple homotopy type if and only if Y can be obtained from X by a sequence of elementary expansions and contractions.

Remark 2. It is known that $\tau(f)$ is unchanged if we subdivide X and Y. However, the topological invariance of $\tau(f)$ is an important open question.

Step 4. Finally, if $(W; M, M')$ is an h-cobordism, we choose triangulations such that the inclusion map $j: M \to W$ is simplicial and define $\tau(W, M) = \tau(j)$ in $Wh(\pi, W)$. (If W is differentiable, we choose the essentially unique differentiable triangulation.)

3. Some further properties of h-cobordisms

We begin this section with a result of Stallings [9].

THEOREM. Suppose $(W; M, M')$ is an h-cobordism with $\dim W \geq 6$. Then $W - M' \approx M \times [0, 1)$.

An easy consequence is the following.

COROLLARY. For any h-cobordism $(W; M, M')$ with $\dim W \geq 6$, $W - (M \cup M') \approx M \times (0, 1)$ and $M \times (0, 1) \approx M' \times (0, 1)$.

To prove the theorem above, let $\tau = \tau(W, M)$ and use Stallings' theorem to find an h-cobordism (W_1, M', M_1) with $\tau(W_1, M') = -\tau$. Now we construct a new h-cobordism $(W \cup_{M'} W_1, M, M_1)$ by attaching W_1 to W along M'. It is easily shown that

$$\begin{aligned}\tau(W \cup_{M'} W_1, M) &= \tau(W, M) + \tau(W_1, M') \\ &= \tau - \tau = 0\end{aligned}$$

so $W \cup_{M'} W_1 \approx M \times I$ and $M_1 \approx M$. In the same way we show that $W_1 \cup_M W \approx M' \times I$.

Now consider the infinite union

$$V = W \cup_{M'} W_1 \cup_M W \cup_{M'} W_1 \cup \cdots .$$

On the one hand,

$$\begin{aligned}V &= W \cup_{M'} (W_1 \cup_M W) \cup_{M'} (W_1 \cup_M W) \cup \cdots \\ &\approx W \cup_{M'} (M' \times I) \cup_{M'} (M' \times I) \cup \cdots \\ &\approx W \cup_{M'} (M' \times [0, 1)) \approx W - M' .\end{aligned}$$

On the other hand,

$$\begin{aligned} V &= (W \cup_{M'} W_1) \cup_M (W \cup_{M'} W_1) \cup \cdots \\ &\approx M \times I \cup_M M \times I \cup \cdots \\ &\approx M \times [0, 1) \end{aligned}$$

and the theorem is proved.

Another result in the same spirit as the theorem above is the following. (See Kwun and Szczarba [4].)

THEOREM. Let $(W; M, M')$ be an h-cobordism with $\dim W \geq 5$ and S^1 the circle. Then

$$W \times S^1 \approx M \times S^1 \times I \quad .$$

In particular, $M \times S^1 \approx M' \times S^1$.

Remark 1. One might conjecture that for any h-cobordism $(W; M, M')$, W is topologically homeomorphic to $M \times I$. Note that a non-trivial h-cobordism $(W; M, M')$ with W homeomorphic to $M \times I$ would provide a counterexample to the hauptvermutung for manifolds and to the topological invariance of torsion.

Remark 2. The question raised in Remark 1 is a special case of the following. Suppose two manifolds have isomorphic interiors and isomorphic boundaries. Are they homeomorphic? Notice that two manifolds may have isomorphic interiors and isomorphic boundaries and not be isomorphic. For example, if $(W; M, M')$ is a non-trivial h-cobordism with $\dim W$ odd and $\pi_1 W$ finite Abelian, $W \cup_{M'} W$ can be shown to be a non-trivial h-cobordism (using Milnor's duality theorem, the fact that "conjugation" in $Wh(\pi)$ is trivial for π finite Abelian, and the fact that $Wh(\pi)$ is torsion free for π finite Abelian (see Milnor [6] and Bass [2])). Thus $W \cup_{M'} W$ and $M \times I$ have isomorphic interiors and boundaries but cannot be isomorphic.

REFERENCES

[1] D. Barden, "The structure of manifolds," Thesis, Cambridge University, 1963.

[2] H. Bass, "The stable structure of quite general linear groups," Bull. Amer. Math. Soc., 70 (1964) pp. 429-433.

[3] G. Higman, "The units of group rings," Proc. London Math. Soc., 46 (1940) pp. 231-248.

[4] K. W. Kwun and R. H. Szczarba, "Product and sum theorems for Whitehead torsion," Ann. of Math., 82 (1965) pp. 183-190.

[5] B. Mazur, "Regular neighborhoods and the theorems of Smale," Ann. of Math., 77 (1963) pp. 232-249.

[6] J. Milnor, "Whitehead torsion," mimeographed notes, Princeton University, 1964.

[7] J. Milnor, <u>h-cobordism</u>, Princeton University Press, 1965.

[8] S. Smale, "On structure of manifolds," Amer. J. Math., 84 (1962) pp. 387-399.

[9] J. Stallings, "On infinite processes leading to differentiability in the complement of a point," <u>Differential and Combinatorial Topology</u>, (to appear), Princeton University Press.

[10] J. H. C. Whitehead, "Simple homotopy types," Amer. J. Math., 72 (1950) pp. 1-57.

ADDITIONAL QUESTIONS ON N-MANIFOLDS

1. (Richard Goodrich) It is known that if K is a knot in E^3, then there exists an arc A whose intersection with K is just its end-points and the resulting pair of simple closed curves are unknotted [1]. Can the same thing be done in higher dimensions?, i.e., let S^2 be a piecewise linear, locally flat embedding of a 2-sphere in E^4. Does there exist a disk, D^2, piecewise linear, locally flat in E^4 with $D^2 \cap S^2 = \text{Bd } D^2$ such that the resulting two spheres bound 3-cells? Is the locally flat condition necessary?

   Reference

   1. S. Kinoshita and M. Terasaka, "On unions of knots," Osaka Math. J., 9, pp. 131-153.

2. (Burt Casler) Is the link of every 1-simplex in a triangulated 5-manifold simply connected?

3. (Burt Casler) Is there a line spanning E^5 so wild that no 3-sphere links the line?

4. (Burt Casler) Does there exist a line in E^5 so wild that every 5-ball containing an interval of the line has a boundary which intersects the line in a Cantor set which is wild with respect to the boundary of the ball?

5. (Burt Casler) If the open star of a 1-simplex in a triangulated 5-manifold is homeomorphic to E^5, will the image of the open 1-simplex (under this homeomorphism) be tame in E^5?

6. (A. C. Connor) Let C be a 4-cell and S its boundary. Suppose that h is a crumpled cube in S such that $\overline{S-h}$ is a real cell. If you sew two copies of C together along h, will the result be a 4-cell?

7. (A. C. Connor) Is every monotone mapping of S^n onto itself essential?

8. (Herman Gluck) Consider the manifold $S^1 \times D^n$ and its boundary $S^1 \times S^{n-1}$, $n \geq 3$. Conjecture: every locally flat $(n-1)$-sphere on $S^1 \times S^{n-1}$ bounds a locally flat n-cell in $S^1 \times D^n$.

10. (Mike Yohe) Bing has described wild 2-spheres which are the fixed point sets of involutions of S^3. Are there wild 3-spheres in S^4 which are the fixed point sets of involutions of S^4?

11. (R. H. Bing) Let A be an arc in E^3 and E^3/A be the decomposition of E^3 whose only non-degenerate element is A. Could $E^1 \times A$ be the fixed point set of an involution of $E^4 = E^1 \times (E^3/A)$? It can if A is either tame or symmetric with respect to some point.

COMPLETELY REGULAR MAPPINGS, FIBER SPACES, THE WEAK BUNDLE PROPERTIES, AND THE GENERALIZED SLICING STRUCTURE PROPERTIES

by

Louis F. McAuley[‡]

Introduction

The various concepts of a fiber space are generalizations of a Cartesian product $B \times F$ where each of B and F is a topological space. Indeed, products are trivial fiber spaces. The projection mapping p from $T = B \times F$ to B is a fiber mapping with fiber F and base space B. In some cases, G is a topological group. However, G is isomorphic to a group of homeomorphisms of F onto F where F is the fiber and G operates transitively on F. The reader may consult Steenrod [8], Hu [4], and Fadell [2, 3] among others for a variety of definitions of fiber spaces and their relationships. It appears that these concepts are aimed towards the development of the so-called <u>exact sequence of a fiber space</u> (<u>or fibering</u>). This sequence is one of the fundamental results in fiber space theory and it is clearly a very powerful tool.

Some definitions of a fiber mapping imply that $p\colon T \to B$ is an open mapping, i.e., $p(U)$ is open relative to $p(T)$ for each open set U in A, while others require reasonable conditions on B and T in order that p be open, e.g., consult the papers [2, 3] by Fadell. Although fibers need not be homeomorphic under the most general concept of a fiber space, this is clearly the case for all fiber bundles and locally trivial fiber spaces where the base space B is connected.

We think that the following conjecture is plausible:

CONJECTURE. Suppose that (T, B, p) is a fiber space such that p has the polyhedral covering homotopy property (PCHP) [cf. 4]. Furthermore, each of T and B is a finite dimensional Peano continuum (locally connected compact metric continuum). If each fiber over a point is

---

‡ The research for this paper was supported in part by NSF GP-4571

homeomorphic to a fixed Peano continuum F, then p has the bundle property (local triviality).

In this paper, we are primarily concerned with the following problem.

Problem. Under what conditions is an open mapping $p\colon T \to B$ a fiber mapping?

This seems to be a rather natural question to raise since fiver mappings are essentially open mappings with rather strong conditions, e.g., the various covering homotopy properties and slicing structure properties.

There are a few answers to this question available to us. In the paper [1], Dyer and Hamstrom have defined the completely regular mapping. And, they have shown that under rather strong conditions (local n-connectedness for the space of homeomorphisms of the inverse of a point onto itself, among others) such a mapping is topologically equivalent to a mapping with the bundle property. We show that a more general class of completely regular mappings have at least one of two weak bundle properties as defined here. Furthermore, we show that certain completely regular mappings have one of two properties which we call generalized slicing structure properties.

Definitions. In this paper, the word mapping means continuous mapping onto unless stated to the contrary. A metric space T is said to be locally n-connected where n is a non-negative integer iff for each integer k, $0 \leq k \leq n$, each point x in T, and each $\varepsilon > 0$, there is a $\delta > 0$ such that each mapping f of a k-sphere S^k into the δ-neighborhood $N_\delta(x)$ of x, there is an extension F of f taking the k+1-disk D^{k+1} into $N_\varepsilon(x)$ [cf. 1]. A mapping p of a metric space T onto a metric space B is homotopically n-regular iff for each integer k, $0 \leq k \leq n$, each x in T and each $\varepsilon > 0$, there exists a $\delta > 0$ such that each mapping $f\colon S^k$ into $N_\delta(x) \cap p^{-1}(b)$ for b in B can be extended to a mapping $F\colon D^{k+1}$ into $N_\varepsilon(x) \cap p^{-1}(b)$ [cf. 1, 7].

Completely Regular Mappings and the Weak Bundle Property I.

The bundle property [4] for fiber maps is rather nice but it is unfortunately too much to expect for open mappings. We define a considerably weaker property which preserves in a sense some of the desirable properties of a bundle or local Cartesian product.

Definition. A mapping $p\colon T \to B$ is said to have the weak bundle property I with respect to a space K iff for each point b in B and mapping g_b of K onto $p^{-1}(b)$, there exists an open set U_b containing b and a mapping $\varphi_{U_b}\colon U_b \times K \to p^{-1}(U_b)$ such that

(1) $p\varphi_{U_b}(u, k) = u$ for (u, k) in $U_b \times K$,

(2) $\varphi_{U_b} | (u, K)$ maps (u, K) onto $p^{-1}(u)$, and

(3) φ_{U_b} is an extension of g_b .

Note: We could require in (2) that (u, K) be mapped into $p^{-1}(u)$ instead of onto $p^{-1}(u)$.

Definition (Dyer and Hamstron) [1]. A mapping $p\colon T \to B$ where each of T and B is a metric space is said to be completely regular iff for each $\varepsilon > 0$ and each point b in B, there is a $\delta > 0$ such that if $x \in B$ and $d(x, b) < \delta$, then there exists a homeomorphism h_{bx} of $p^{-1}(b)$ onto $p^{-1}(x)$ which moves no point as much as ε.

In the theorem which follows, we show that a class of completely regular mappings has the weak bundle property I.

THEOREM 1. Suppose that $p\colon T \to B$, each T and B is a metric space, T is complete, covering dimension $B \leq n+1$, and p is completely regular. Furthermore, there is a metric space K such that for each b in B, there is a mapping of K onto $p^{-1}(b)$ and that the space G_b of all such mappings is locally n-connected (LC^n). Then p has the weak bundle property I w.r.t. K. [If K is homeomorphic to $p^{-1}(b)$ and G_b is the space of all homeomorphisms of K onto $p^{-1}(b)$, then p has the bundle property — Dyer and Hamstron.]

The basic ideas for the proof of this theorem came from the paper [1]. There are enough subtle differences so that our presentation of the details seems justifiable. The proof depends (as those proofs in [1] do) on an important and powerful selection theorem of E. A. Michael [5, 6] which we state below.

THEOREM M. If each of A and B is a metric space, A is complete, covering dimension of $B \leq n+1$, Z is a closed subset of B, F is a mapping of A onto B such that the collection of inverses under F is lower semicontinuous (defined below) and equi-LC^n (as defined below), and f is a mapping of Z into A such that for z in Z, $f(z) \in F^{-1}(z)$, then there is a neighborhood U of Z in B such that f can be extended to a mapping f^* of U into A such that for $b \in U$, $f^*(b) \in F^{-1}(b)$. If each inverse under F has the property that its homotopy groups of order $\leq n$ vanish, then U may be taken to be the space B.

Definition. A collection G of closed point sets filling a metric space X (i.e., the union of the elements of G is X) is said to be equi-LC^n iff for each $\varepsilon > 0$, g in G, and $x \in g$, there is a $\delta > 0$ such that if $h \in G$ and f is a mapping of a k-sphere S^k, $0 \leq k \leq n$ into $h \cap N_\delta(x)$, then there is an extension F of f to the k+1-disk D^{k+1}, into $h \cap N_\varepsilon(x)$, [cf.1].

We shall assume that T has a bounded complete metric d. Furthermore, the metric for G_b is understood to be as follows: For f, g in G_b, $D(f, g) = \max\{d[f(x), g(x)]\}$ for x in K. We shall also use d to denote a metric for B. No confusion should arise.

Notation. Let G denote the collection of all G_b for b in B and let G^* denote the space of those mappings of K into T which is the union of the elements of G. The metric for this space is defined as above.

LEMMA 1.1. The metric space G^* is complete.

Proof: Suppose that $\{z_i\}$ is a Cauchy sequence in G^*. Since T is complete and the space Z of all mappings of K into T is complete, $\{z_i\} \to z \in Z$. Now, z_i is a mapping of K onto $p^{-1}(b_i)$. Also, $\{b_i\} \to b$ in B. It follows from the hypotehsis that if $\varepsilon > 0$, there exists m so that if $n > m$, then there is a mapping $f_{b_n b}$ of $p^{-1}(b_n)$ onto $p^{-1}(b)$ which moves no point as much as $\varepsilon/3$ and $d(z, z_n) < \varepsilon/3$. Observe that $g_n = f_{b_n b} z_n$ is a mapping of K onto $p^{-1}(b)$ such that $d(g_n, z_n) < 2\varepsilon/3$ and $d(g_n, z) < \varepsilon$. Therefore $z \in G_b$ and G^* is complete.

LEMMA 1.2. The collection G is equi-LC^n.

Proof: Suppose that $g \in G_b$ and $\varepsilon > 0$. Since G_b is LC^n, there is $\delta_1 > 0$ such that each mapping r of S^k, $0 \leq k \leq n$, into $G_b \cap N_{\delta_1}(g)$ can be extended to a mapping R of D^{k+1} into $G_b \cap N_{\varepsilon/2}(g)$. Assume that $\delta_1 < \varepsilon/2$. Also, there is $\alpha > 0$ such that if $d(b, c) < \alpha$, $c \in B$, then there is a homeomorphism h_{cb} of $p^{-1}(c)$ onto $p^{-1}(b)$ which moves no point as much as $\delta_1/2$. Choose δ, $0 < \delta < \delta_1/2$, such that if $y \in G_c \cap N_\delta(g)$, then $d(b, c) < \alpha$.

Now, let $\varphi\colon S^k \to G_c \cap N_\delta(g)$. We wish to show that φ can be extended to $\Phi\colon D^{k+1}$ into $G_c \cap N_\varepsilon(g)$. Note that h_{cb} determines a 1-1 mapping H_{cb} of G_c into G_b as follows: For e in G_c, $H_{cb}(e) = h_{cb}e$. And, $H_{cb}|G_c \cap N_\delta(g)$ maps $G_c \cap N_\delta(g)$ into $G_b \cap N_{\delta_1}(g)$ since h_{cb} maps $p^{-1}(c)$ homeomorphically onto $p^{-1}(b)$ without moving any point as much as $\delta_1/2$. Furthermore, $r = [H_{cb}|\varphi(S^k)]\varphi$ maps S^k into $G_b \cap N_{\delta_1}(g)$ and can be extended to a mapping R of D^{k+1} into $G_b \cap N_{\varepsilon/2}(g)$.

The mapping h_{cb}^{-1} determines a 1-1 mapping H_{bc} of G_b into G_c defined in the manner indicated above. Each element s in $G_b \cap N_{\varepsilon/2}(g)$ maps K onto $p^{-1}(b)$, h_{cb}^{-1} maps $p^{-1}(b)$ homeomorphically onto $p^{-1}(c)$, and h_{cb}^{-1} moves no point as much as $\delta_1/2 < \varepsilon/2$. Furthermore,

$$\Phi = [H_{bc}|G_b \cap N_{\varepsilon/2}(g)]R \text{ maps } D^{k+1} \text{ into } G_c \cap N_\varepsilon(g)$$

and agrees with φ on S^k, the boundary of D^{k+1}. The lemma is proved.

LEMMA 1.3. The collection G is lower semicontinuous. That is, $\{b_i\} \to b$ where b_i, $b \in B$, then G_b is in the closure of $\cup G_{b_i}$.

Proof: There exists a sequence $\{\delta_i\} \to 0$ such that if $d(b, b_n) < \delta_i$, then there exist mappings (in this case homeomorphisms although the lemma may be proved using mappings with similar properties)

$$f_{bb_n}: p^{-1}(b) \to p^{-1}(b_n) \quad \text{and} \quad f_{b_n b}: p^{-1}(b_n) \to p^{-1}(b)$$

which move no point as much as $1/i$. Let g denote an element of G_b. That is, $g: K \to p^{-1}(b)$. And, $f_{bb_n} \cdot g$ is an element of G_{b_n}. Now, $d(g, f_{bb_n} g) < 1/i$. It follows that g is in the closure of $\cup G_{b_n}$.

Proof of Theorem 1. The mapping F defined by $F(G_b) = b$ for each b in B is a mapping of G^* onto B and satisfies the hypothesis of Theorem M. Now, for b in B, select g_b in G_b and let $f(b) = g_b$. The closed set $\{b\}$ is the set Z of Theorem M while $f: Z \to T$ (where T is A). Therefore, there is by Theorem M, an open set U_b containing b and a mapping $g_{U_b}: U_b \to G^*$ which is a continuous extension of f such that for $u \in U_b$, $g_{U_b}(u) \in G_u$.

For (u, k) in $U_b \times K$, let $\varphi_{U_b}(u, k) = (u, [g_{U_b}(u)](k))$. It follows that φ_{U_b} maps $U_b \times K$ onto $p^{-1}(u)$, φ_{U_b} is an extension of g_b, and indeed p has the weak bundle property I w.r.t. K.

COROLLARY (Theorem of Dyer and Hamstrom). If K is homeomorphic to $p^{-1}(b)$ for each b in B and the space $H(K)$ of all homeomorphisms of K onto K is LC^n, then p has the bundle property.

Question. For what spaces K is the space of all continuous mappings of K onto LC^n? If $p: T \to B$ is a completely regular mapping, then for what spaces K, if any, is the space of all continuous mappings of K onto $p^{-1}(b)$ LC^n for $b \in B$?

Remark. We could have used into mappings in Theorem 1 instead of onto mappings. The result would be a mapping $\varphi_{U_b}: U \times K$ into $p^{-1}(U)$. Perhaps, this would be more useful since the space of all continuous mappings of a space K into K [or into $p^{-1}(b)$] may be LC^n for some spaces K whereas the space of all homeomorphisms of K onto [or into $p^{-1}(b)$] may fail to be LC^n.

Completely Regular Mappings and the Weak Bundle Property II.

The concept of a mapping $p: T \to B$ with the WBP II is useful in studying light open mappings p. One may consider a space K such as the interval $I = [0, 1]$ and the space of all mappings f of $p^{-1}(b)$ into I where $p^{-1}(b)$ is totally dosconnected. Such a space G_b is LC^n for all n. Some interesting results are obtained when $p^{-1}(b)$ is homeomorphic to the Cantor set. Perhaps some progress can be made towards settling the outstanding question of whether a p-adic group can act on an n-manifold through the use of the WBP II.

Definition. A mapping $p: T \to B$ is said to have the weak bundle property II with respect to a space K if for each b in B and each mapping g_b: $p^{-1}(b)$ into K, there exists an open set U_b containing b and a mapping $\varphi_{U_b}: p^{-1}(U_b)$ into $U_b \times K$ such that

(1) $p\,\varphi_{U_b}^{-1}(u, k) = u$ for each $(u, k) \in U_b \times K$,

(2) $\varphi_{U_b}|p^{-1}(u)$ maps $p^{-1}(u)$ into (u, K), and

(3) φ_{U_b} is a continuous extension of g_b.

That is, φ_{U_b} maps $p^{-1}(U_b)$ into $U_b \times K$, φ_{U_b} extends g_b, and $p|p^{-1}(U_b) = \pi\,\varphi_{U_b}$ where π is the projection mapping of $U_b \times K$ onto U_b.

THEOREM 2. Suppose that $p: T \to B$, each of T and B is a metric space, T is complete, covering dimension of $B \leq n+1$, and p is completely regular. Furthermore, K is a complete metric space such that the space G_b of all mappings of $p^{-1}(b)$ into K is LC^n for each b in B. Then p has the weak bundle property II w.r.t. K.

Here K is given. Obviously, there exist such spaces K. For example, let K be a point space. But, this is a trivial case and uninteresting.

Question. For what interesting spaces K is the space G_b of all mappings of $p^{-1}(b)$ into K LC^n?

A proof of Theorem 2 follows the same pattern as that given for Theorem 1 provided that there is a complete metric for G^* where G is the collection of all G_b for b in B. This is easy when G_b denotes the space of all mappings of K into $p^{-1}(b)$. It is more complicated in the cases (as above) when G_b is the space of all mappings of $p^{-1}(b)$ into K. However, it is possible. In fact, this mapping space turns out to be both interesting and useful. For details, see the paper, "The existence of a complete metric for a special mapping space" in this volume.

Generalized Slicing Structure Properties and Completely Regular Mappings

Consider the definition below of the slicing structure property as given by Hu [4].

Definition. Suppose that $p: T \to B$ is a mapping. We say that p has the slicing structure property iff there exists an open covering $\mathcal{Q}$ of B such that for each element U of $\mathcal{Q}$, there is a mapping φ_U which maps $U \times p^{-1}(U) \to p^{-1}(U)$ such that

(1) $p\varphi_U(b, x) = b$ and

(2) $\varphi_U(p(x), x) = x$.

We say that T is a sliced fiber space over B relative to p.

Another way to think of a sliced fiber space T over B relative to p is as follows: For each point $b \in B$, there is an open set U_b containing b and a mapping $\varphi_{U_b}: p^{-1}(U_b) \to p^{-1}(b)$ such that

(a) $\varphi_{U_b}|p^{-1}(b)$ is the identity mapping and

(b) the collection $\{\varphi_{U_b}\}$ is continuous in the sense that if $\{b_i\} \to b$, then $\{\varphi_{U_{b_i}}|U_b\} \to \varphi_{U_b}$.

One may generalize the notion of a sliced fiber space in the following way.

Definition. We shall say that $p: T \to B$ has the generalized slicing structure property (c) w.r.t. a space K denoted by GSSP(c) iff for each point b in B there is an open set U_b containing b and a mapping $\varphi_{U_b}: p^{-1}(U_b)$ into K such that

(a) $\varphi_{U_b}|p^{-1}(b)$ is a continuous mapping of $p^{-1}(b)$ onto K and

(b) the collection $\{\varphi_{U_b}\}$ is continuous.

If we require that in (a) above, $\varphi_{U_b}|p^{-1}(b)$ is a homeomorphism of $p^{-1}(b)$ onto K for each b in B, then we shall say that p has the generalized slicing structure (h) with respect to the space K denoted by GSSP(h). If this homeomorphism is the identity for each b in B, then we have the usual slicing structure property.

The following theorems show that under certain conditions, a completely regular mapping has either the GSSP(c) or GSSP(h) or SSP.

THEOREM 3. Suppose that $p\colon T \to B$ is a completely regular mapping, T is a complete metric space, and B is a metric space with covering dimension of $B \leq n+1$. Furthermore, there is a metric space K, an open covering W of B, and a continuous collection H of homeomorphisms such that (1) for w in W, and $b \in w$, there is h_w in H such that h_w maps K onto $p^{-1}(w)$ and (2) the space of all mappings f, h_w where f is a mapping of $p^{-1}(w)$ onto $p^{-1}(b)$ such that $f|p^{-1}(b)$ maps $p^{-1}(b)$ onto $p^{-1}(b)$ is LC^n, then p has the GSSP(c).

THEOREM 4. Suppose that $p\colon T \to B$ is a completely regular mapping, T is a complete metric space, and B is a metric space with covering dimension of $B \leq n+1$. Furthermore, there is an open covering W of B such that if $b \in w \in W$, then the space G_b of all mappings f of $p^{-1}(w)$ into $p^{-1}(b)$ such that

$$f|p^{-1}(b) \text{ maps } p^{-1}(b) \begin{cases} \text{onto } p^{-1}(b) \\ \text{homeomorphically onto } p^{-1}(b) \\ \text{identically onto } p^{-1}(b) \end{cases}$$

is LC^n. Then p has the

$$\left\{\begin{matrix} \text{GSSP(c)} \\ \text{GSSP(h)} \\ \text{SSP.} \end{matrix}\right\} \quad \text{with respect to } K \cong \{p^{-1}(b)\}.$$

Proofs of Theorems 3 and 4 may be obtained by mimicing the argument for Theorem 1.

Remarks. An obvious generalization of the concept of complete regularity is that of regularity which is as follows:

A mapping $p\colon T \to B$ where each of T and B is metric is said to be regular iff for each $\varepsilon > 0$ and each b in B, there is a $\delta > 0$ such that if $x \in B$ and $d(x, b) < \delta$, then there exists a mapping f_{xb} of $p^{-1}(x)$ onto $p^{-1}(b)$ which moves no point as much as ε. If f_{xb} (in each case) is a homeomorphism, then p is completely regular [1] as used in this paper. We could require in the definition above that f_{xb} maps $p^{-1}(x)$ into $p^{-1}(b)$ instead of onto and consider the consequences of this change.

Question. Is it possible to prove theorems analogous to those above where the complete regularity of the mapping $P\colon T \to B$ is replaced by regularity? (Yes, McAuley and Tulley)

If we attempt to use the techniques of Dyer and Hamstrom, then we must show that the collection G (as defined in the various theorems) is equi-LC^n. The proofs in [1] and those either given or indicated above depend on the complete regularity of p.

REFERENCES

[1] Dyer, Eldon and Hamstrom, M.-E., "Completely regular mappings," Fund. Math., vol. 45 (1957) pp. 103-118.

[2] Fadell, E. R.,"On fiber spaces," TAMS, vol. 90 (1959) pp. 1-14.

[3] ______, unpublished notes.

[4] Hu, S-T-. Homotopy Theory, Academic Press, New York, N. Y., 1959.

[5] Michael, E. A., "Continuous selections, I, II and III," Ann. Math., vol. 63 (1956) pp. 361-382; vol. 64 (1956) pp. 562-580; vol. 65 (1957) pp. 357-390.

[6] ______, "Selection Theorems for continuous functions," Proc. Int. Congress Math., Amsterdam, 1955, vol. II, 241.

[7] Puckett, W. T., Jr., "Regular transformations," Duke Jour., vol. 6 (1940) pp. 80-88.

[8] Steenrod, N., The Topology of Fibre Bundles, Princeton University Press, Princeton, N. J., 1951.

[9] Whyburn, G. T., "On sequences and limiting sets," Fund. Math. vol. 25 (1935) pp. 408-426

RUTGERS UNIVERSITY

FIBER SPACES AND n-REGULARITY

by

L. F. McAuley* and P. A. Tulley

Introduction

There is a theorem due to E. A. Michael [4] which, roughly stated, says that an open and closed mapping p (continuous) from a complete metric space T is a Serre fiber mapping provided that p is uniformly n-regular (in the homotopy sense) for each integer $n \geq 0$. On the other hand, Raymond [5] has shown that if $p: T \to B$ is a fiber mapping in the sense of Hu [2], i.e., p has the slicing structure property, and T is locally n-connected, then p is n-regular. This is an immediate consequence of the slicing structure property, a condition much stronger than the polyhedral covering homotopy property (Serre fibration). One is led, quite naturally, to the following question.

Question. Suppose that $p: T \to B$ has the polyhedral covering homotopy property (PCHP)[2]. Under what conditions is p n-regular?

We show that if each of T and B is metric, T is locally n-connected, B is locally n+1-connected, and p is a mapping of T onto B which has the PCHP, then p is n-regular. In order to do this we show that certain small homotopies can be lifted to small homotopies.

Definitions. A metric space T is said to be locally n-connected (LC^n) if and only if for each x in T, $\varepsilon > 0$, and integer k with $0 \leq k \leq n$, there is a $\delta > 0$ such that for each mapping f of a k-sphere S^k into the δ-neighborhood $N_\delta(x)$, there is an extension F of f mapping the k+1-cell D^{k+1} (where S^k is the boundary of D^{k+1}) into $N_\varepsilon(x)$ [1]. A mapping p of a metric space T onto a metric space B is homotopically n-regular if and only if for each x in T, $\varepsilon > 0$ and integer k with $0 \leq k \leq n$, there is a $\delta > 0$ such that if f is a mapping of S^k into $N_\delta(x) \cap p^{-1}(b)$ for b in B, then f has an extension F mapping D^{k+1} into $N_\varepsilon(x) \cap p^{-1}(b)$ [5]. A mapping p of T into B is said to have the covering homotopy property (CHP) with respect to a space X if and only if, given a mapping f of X into B, a lift-

* Research for this paper was supported in part by NSF GP-4571.

ing f^* mapping X into T with $pf^* = f$, and a homotopy H taking $X \times I$ into B such that $H(x, 0) = f(x)$ for each x in X, there is a homotopy H^* taking $X \times I$ into T such that $H^*(x, 0) = f^*(x)$ for each x in X and $pH^* = H$ [2]. In the situation just described, we shall say that H^* is a lifting of H relative to f^*. If p has the CHP for each polyhedron (finite), then p is said to have the polyhedral covering homotopy property (PCHP) or, equivalently, (T, p, B) is said to be a Serre fiber space.

The following fact should be noted. Suppose that $p\colon T \to B$, p has the CHP with respect to X, and B is a metric space. Furthermore, suppose that f, f^*, and H are as above. Then there is a lifting H^* of H relative to f^* such that if, for x in X, $H(x, t)$ is independent of t, then $H^*(x, t)$ is also independent of t, i.e., if x in X has the property that $H(x, t) = H(x, 0)$ for each t in I, then $H^*(x, t) = H(x, 0)$ for each t in I. This fact is crucial to our proof of Theorem 1. It can be proved by an argument similar to that given in [3] where it is shown that any Hurewicz fiber space with a metric base space is regular, i.e., has a lifting function taking constant paths to constant paths.

Lifting Small Homotopies to Small Homotopies

Suppose that $p\colon T \to B$ has the CHP with respect to a space X and that each of T and B is metric. We shall say that small homotopies of X can be lifted to small homotopies if and only if for each $\varepsilon > 0$ and each e in T, there is a $\delta > 0$ such that if H is a mapping of $X \times I$ into $N_\delta(p(e))$ and f^* is a mapping of X into $N_\delta(e)$ with $pf^*(x) = H(x, 0)$ for x in X, then there is a lifting H^* of H relative to f^* such that H^* maps $X \times I$ into $N_\varepsilon(e)$. Theorem 1 states that small homotopies of an n-cell can be lifted to small homotopies for certain Serre fiber spaces. This is the fact needed in our proof of Theorem 2. However, the restriction that X ba an n-cell is not necessary. After proving Theorem 1 we will indicate how it can be generalized to the situation where X is any polyhedron.

THEOREM 1 (Tulley). Suppose that $p\colon T \to B$ has the PCHP, each of T and B is metric, T is LC^0 and X is an n-cell. Then small homotopies of X can be lifted to small homotopies.

Proof: Suppose that the theorem is not true. Then, since T and B are metric, there are sequences $\{H_i\}$ and $\{g_i\}$ of mappings such that:

(1) for each i, $H_i\colon X \times I \to B$, $g_i\colon X \to T$, and $pg_i(x) = H_i(x, 0)$ for each x in X,

(2) $\{g_i(X)\} \to e$ for some e in T, and

(3) $\{H_i(X \times I^*)\} \to b = p(e)$.

Also, there is no sequence $\{H_i^*\}$ such that H_i^* is a lifting of H_i relative to g_i for each i and $\{H_i^*(X \times I)\} \to e$.

Let Z denote $C(X)$, the cone over X, i.e., the quotient space obtained from $X \times I$ by collapsing $X \times \{0\}$ to a point v, the vertex of the cone. We identity $Z - \{v\}$ with $X \times (0, 1]$. Let $X_i = \{(x, s) \in Z | s = 1/i\}$ for each positive integer i. Choose any x_0 in X and let $Y_i = X_{i+1} \cup X_i \cup (\{x_0\} \times [1/i + 1, 1/i])$ for each i and let

$$A = \left(\bigcup_{i=1}^{\infty} Y_i\right) \cup \{v\}.$$

Each Y_i is contractible and therefore is an absolute retract. It follows that there is a retraction $r\colon Z \to A$.

We define a mapping g of Z into T as follows: Let $g(x, 1/i) = g_i(x)$ for each i and each x in X. This defines $g| \cup_{i=1}^{\infty} X_i$. Since T is LC^0, for sufficiently large i we may extend g to Y_i in such a way that $\{g(Y_i)\} \to e$. We assume without loss of generality that these extensions can be made for each i. Letting $g(v) = e$ we complete the definition of $g|A$ and then we define g by $g = (g|A)r$.

We define a mapping H of $Z \times I$ into B as follows: First we define $K\colon (\cup X_i) \times I \cup Z \times \{0\} \to B$ by $K((x, s), 0) = pg(x, s)$ for $((x, s), 0)$ for x in X and s in $(0, 1]$, $K(v, 0) = b$ and $K((x, 1/i), t) = H_i(x, t)$ for $((x, 1/i), t)$ in $(\cup_{i=1}^{\infty} X_i) \times I$. Now, foreach i, we let r_i be a retraction of $[1/i+1, 1/i] \times I$ onto

$$(\{1/i+1\} \times I) \cup ([1/i+1, 1/i] \times \{0\}) \cup (\{1/i\} \times I)$$

and for any (x, t) in $(0, 1] \times I$ let (s', t') be the point $r_i(s, t)$ for any i for which $1/i+1 \leq s \leq 1/i$. We define $H((x, s), t) = K((x, s'), t')$ for $((x, s), t)$ in $(Z - \{v\}) \times I$ and $H(v, t) = b$ for each t in I. It is easy to see that H is a (continuous) mapping of $Z \times I$ into B and that $pg(z) = H(z, 0)$ for each z in Z.

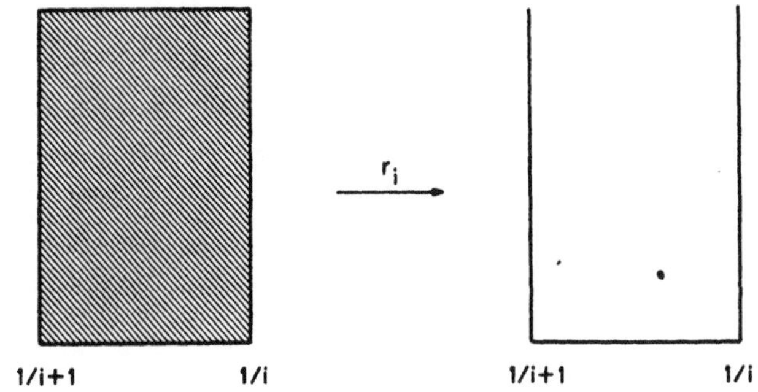

Since the PCHP holds for $p\colon T \to B$, there is a mapping H^* of $Z \times I$ into T such that $H^*(z, 0) = g(z)$ for each z in Z and $pH^* = H$. Furthermore, since B is metric, we can suppose that $H^*(v, t) = e$ for each t in I.

Letting $H_i^*(x, t) = H^*((x, 1/i), t)$ we get a sequence $\{H_i^*\}$ of homotopies mapping $X \times I$ into T such that $H_i(x, 0) = g_i(x)$ for each x in X, $pH_i^* = H_i$, and, by the continuity of H, $\{H_i(X \times I)\} \to e$. This is a contradiction and the theorem is proved.

Remarks. With Theorem 1 and the techniques used in [2] (pp. 63-64) to show the equivalence of various lifting properties for Serre fiber spaces, the following stronger theorem can be proved. One must be careful to obtain sufficient "smallness" at each stage of the proof but this is easy to do.

THEOREM 1'. Suppose that $p\colon T \to B$ has the PCHP, X is a polyhedron, each of T and B is metric, and T is LC^0. Then small homotopies of X can be lifted to small homotopies.

Also, it should be noted that if (T, p, B) is a Hurewicz fiber space, i.e., p has the CHP for every space, and each of T and B is metric, then small homotopies of any space X can be lifted to small homotopies. The requirement that T be LC^0 is not needed. In fact, a proof is easily obtained by following the procedure used in the proof of Theorem with $(\bigcup_{i=1}^{\infty} X_i) \cup \{v\}$ playing the role of Z. In this case, the appropriate mappings g and H are trivial to define.

A Fiber Mapping which is n-Regular

THEOREM 2 (McAuley). Suppose that $p\colon T \to B$ has the PCHP, each of T and B is metric, T is LC^n and B is LC^{n+1}. Then p is homotopically n-regular.

Proof: Suppose that $e \in T$, $\varepsilon > 0$, $b = p(e)$, and $0 \le k \le n$. Then there exist δ_i for $i = 1,2,3$, and 4 such that:

(1) if g is a mapping of D^{k+1} into $N_{\delta_1}(e)$ and H is a homotopy taking $D^{k+1} \times I$ into $N_{\delta_1}(b)$ such that $pg(x) = H(x, 0)$ for each x in D^{k+1}, then there is a lifting H^* of H relative to g such that H^* maps $D^{k+1} \times I$ into $N_\varepsilon(e)$; (we use Theorem 1 to get δ_1)

(2) any mapping of S^{k+1} into $N_{\delta_2}(b)$ can be extended to a mapping of D^{k+2} into $N_{\delta_1}(b)$;

(3) $N_2(b) \supset p(N_{\delta_3}(e))$; and

(4) any mapping of S^k into $N_{\delta_4}(e)$ can be extended to a mapping of D^{k+1} into $N_{\delta_1}(e) \cap N_{\delta_3}(e)$.

Let $\delta = \delta_4$ and consider $f\colon S^k \to N_\delta(e) \cap p^{-1}(b')$ for some b' in B. Let F be an extension of f mapping D^{k+1} into $N_{\delta_3}(e) \cap N_{\delta_1}(e)$. Clearly, pF is a mapping of D^{k+1} into $N_{\delta_2}(b)$ such that $pF(S^k) = b'$. We consider S^{k+1} as D^{k+1} with S^k identified to a point z. Then pF yields a mapping g of S^{k+1} into $N_{\delta_2}(b)$ such that $g(z) = b'$. Let G be an extension of g mapping D^{k+1} into $N_{\delta_1}(b)$. By shringkin S^{k+1} to the point z in D^{k+2} while keeping z fixed and following this shrinking by the mapping G, we obtain a homotopy taking $S^{k+1} \times I$ into $N_{\delta_1}(b)$. This induces a homotopy H taking D^{k+1} into $N_{\delta_1}(b)$ with the properties that $H(x, 0) = pF(x)$ for each x in D^{k+1} and $H((S^k \times I) \cup (D^{k+1} \times \{1\})) = b'$. Now, there is a lifting H* of H relative to F mapping $D^{k+1} \times I$ into $N_\varepsilon(e)$. It follows that $H^*(x, 0) = f(x)$ for x in S^k and that

$$p^{-1}(b') \cap N_\varepsilon(e) \quad H^*((S^k \times I) \cup (D^{k+1} \times \{1\})).$$

Let $A = (S^k \times I) \cup (D^{k+1} \times \{1\})$ and let $\hat{F} = F|A$. Since A is a k+1-cell, it is clear that $\hat{F}$ yields a mapping of D^{k+1} into $p^{-1}(b') \cap N_\varepsilon(e)$ which extends f. This completes the proof.

<u>Remark</u>. If p is homotopically n-regular, then each fiber is LC^n. Thus, Theorem 2 implies a local result analogous to the well-known fact that if $p\colon T \to B$ has the PCHP, T is n-connected and B is n+1-connected then each fiber is n-connected.

REFERENCES

[1] Dyer, E. and Hamstrom M.-E., "Completely regular mappings," Fund. Math., vol. 45 (1957) pp. 103-118.

[2] Hu, S.-T., Homotopy Theory, Academic Press, New York, N. Y., 1959.

[3] Hurewicz, W., "On the concept of fiber space," Proc. Nat. Acad. Sci., U.S.A., vol. 41(1955) pp. 956-961.

[4] Michael, E. A., "Continuous selections III," Annals of Math., vol. 65 (1957) pp. 357-390.

[5] Raymond, Frank, "Local triviality for Hurewicz fiberings of manifolds," Topology, vol. 3 (1965) pp. 43-57.

[6] Serre, J.-P., "Homologie singuliere des espaces fibres," Annals of Math., vol. 54(1951) pp. 425-505.

RUTGERS UNIVERSITY

FIBER SPACES WITH TOTALLY PATHWISE DISCONNECTED FIBERS*

by

Gerald S. Ungar**

Two outstanding problems in topology involving light open mappings are as follows:

I. Does there exist a light open mapping of a manifold onto a manifold such that the inverse of some point is uncountable?

II. Is there a finite to one open mapping f which is not the identity mapping of the n-cube I^n onto itself such that f is the identity on the boundary of I^n (bdry I^n) and $f^{-1}f(\text{bdry } I^n) = \text{bdry } I^n$?

It can be shown (Corollary 1 of Theorem 3) that there are no Serre fibrations with the properties of either I or II. Hence, a method of attacking these problems might be the following: assume that there exists such a mapping and prove that it must be a fibration. The object of this paper is to begin a study of conditions on light maps so that they must be fiber maps. For references see [1], [3], and [7].

The following are some definitions which will be needed. In all cases p is a map from a topological space T onto a topological space B.

1) p is a-<u>light</u> if $p^{-1}(b)$ contains no arcs for every b in B.

2) p <u>has weak local cross sections at every point</u> if given any b in B and any y in $p^{-1}(b)$ there exists a neighborhood U_y of b and a map φ_{U_y} of U_y into T such that $\varphi_{U_y}(b) = y$ and $p\varphi_{U_y}$ is the identity on U_y.

3) p <u>has strong local cross sections at every point</u> if given any b in B there exists a neighborhood U of b such that if y is in

* The author would like to express his gratitude to Professor McAuley for his interest, encouragement, advice and assistance.

** The author was partially supported by an NSF Academic Year Extension.

$p^{-1}(U)$ there exists a map φ_y: $U \to T$ such that $p\varphi_y$ is the identity on U and y is in $\varphi_y(U)$.

4) If Σ is a class of topological spaces; then p is said to have the Σ covering homotopy property (ΣCHP) if given any map f of a space $X \in \Sigma$ into T and a homotopy $H: X \times I \to B$ such that $H(x, 0) = pf(x)$ then there exists a homotopy $K: X \times I \to T$ such that $K(x, 0) = f(x)$ and $pK(x, t) = H(x, t)$.

We will use the following notation. If p has the ΣCHP and

a) Σ is the class of all topological spaces, then p has the ACHP (absolute covering homotopy property, see Hu [4]).
b) Σ is the class consisting of the unit interval, then p has the ICHP.
c) Σ is the class of compact locally arcwise connected spaces, then p has the CLACHP.
d) Σ is the class of finite polyhedra, then p has the PCHP.

If p has the PCHP, then (T, p, B) is called a fiber space in the sense of Serre.

e) Σ is the class consisting of the sets $p^{-1}(b)$ for every $b \in B$ then p has the FCHP.
f) Σ is the class consisting of a one point set, then p has the 0-CHP (i.e., paths can be lifted).

If the mapping K in the previous definition is always unique then p has the uΣCHP.

The first theorem along the lines mentioned above is the following:

THEOREM 1: The following are equivalent:

A) If σ is a loop in B which is homotopic to a constant and $y \in p^{-1}\sigma(0)$, then there is a lifting τ of σ such that $\tau(0) = y$ and any lifting of σ is a loop.

B) If σ is a loop in B which is homotopic to a constant, then there exists a unique loop τ which covers σ such that $\tau(0) = y$.

C) If α is a path in B and $y \in p^{-1}\alpha(0)$, then there exists a path β in T which covers α such that $\beta(0) = y$. Also if α and γ are homotopic paths in B and β and δ are liftings of α and γ respectively, such that $\beta(0) = \delta(0)$, then $\beta(1) = \delta(1)$.

D) Same as C except for the fact that β is assumed to be unique.

E) p is a-light and has the ICHP.

F) p is a-light and has the PCHP.

It should be noted that the first condition of C is equivalent to the 0-CHP and that the first condition of D is equivalent to the u0-CHP. This gives rise to the following conjecture which is easily seen to be true if β is a 1-dimensional complex.

CONJECTURE. The mapping p has the u0-CHP iff p is a-light and has the PCHP.

If it is known a-priori that p has the 0-CHP, then conditions A and C can be weakened to read as follows:

A') If σ is a loop in B which is homotopic to a constant, then every lifting of σ is a loop.

C') If α and γ are homotopic paths in B and β and δ are liftings of α and γ, respectively, such that $\beta(0) = \delta(1)$, then $\beta(1) = \delta(1)$.

There is a theorem by Floyd [2] which is as follows:

THEOREM (Floyd). If p is light open, T and B are compact, connected, metric spaces, then p has the 0-CHP. It can also be shown that if p has strong local cross sections at every point, then p has the 0-CHP. This is not true if it is just assumed that p has weak local cross sections. The following example shows this: Let $T = \{(x, y) \in E^2 | y = 0 \text{ and } 0 \le x \le 1 \text{ or } y = 1 \text{ and } 0 \le x < \frac{1}{2}\}$, let $B = I$ the unit interval and define $p: T \to B$ by $p(x, y) = x$. It would be interesting to know when the 0-CHP implies that p has strong (or even weak) local cross sections at every point. In general it does not. Keldys [6] has reported an example of a light open map of a one-dimensional Peano continuum onto a square. However, suppose that either T is a manifold or that p does not raise dimension, (or even suppose that p is a fibration), then does the 0-CHP imply that p has local sections of some kind. Theorem 3 gives a partial result if p is a fibration. Along these same lines, the following theorem is obtained.

THEOREM 2. If p has weak local cross sections at every point and p has the u0-CHP then p has the CLACHP.

This theorem suggests the following question:

Is the CLACHP equivalent to the PCHP? If not, when is it the case?

Theorems 2 and 3 give partial answers to some of the questions which were raised above. However, the author has not been able to show that these are the best possible.

THEOREM 3. If p is a-light and has the PCHP and B is first countable, locally arcwise connected and semi-locally simply connected, then p has strong local cross sections at every point.

COROLLARY 1. If T, p and B are as in the theorem above, T is locally arcwise connected, and B is arcwise connected then p has the bundle property (B.P.).

COROLLARY 2. If T, p and B are as in the theorem above and $p^{-1}(b)$ is compact for all $b \in B$, B is locally compact, arcwise connected, and p has the FCHP, then p has the B.P.

The following question can be raised with regard to Corollary 2. When do the PCHP and FCHP imply the B.P.? In general they do not since there are one to one fiber maps which are not homeomorphisms, and there is an example [5] of a map which has the ACHP but does not have the B.P.

The following example shows that in general some condition such as the FCHP must be assumed to obtain the B.P.

Example: Let T be the following subset of the plane:

$$T = \{(x, y) \mid y = x + \frac{1}{n}, \ n = 1,2,\ldots \text{ and } 0 \leq x \leq 1\}$$
$$\cup \{(x, y) \mid y = 0 \text{ and } 0 \leq x \leq 1\}$$
$$\cup \{(x, y) \mid y = -\frac{1}{2^n}, \ n=1,2,\ldots \text{ and } \frac{1}{2^n} \leq x \leq 1\}$$
$$\cup \{(x, y) \mid y = 2n^{-1}x - \frac{2^{n-1}+1}{2^n}, \ n = 1,2,\ldots \text{ and } 0 \leq x \leq \frac{1}{2^n}\}$$

Let B be the unit interval and define $p\colon T \to B$ as follows $p(x, y) = x$. The map p has cross sections at every point, p has the ICHP, and by Theorem 2, p has the CLACHP. However, p does not have the FCHP or B.P. even though B is the unit interval and all fibers are homeomorphic.

The proof of these and other related theorems will be published in detail elsewhere.

REFERENCES

[1] F. E. Browder, "Covering spaces, fiber spaces, and local homeomorphisms," Duke Math. J., vol. 71 (1954) pp. 329-336.

[2] E. E. Floyd, "Some characterizations of interior maps," Annals of Math., 51 (1950) pp. 571-575.

[3] J. S. Griffen, "Theorems on fiber spaces," Duke Math. J., 20 (1953) pp. 621-628.

[4] S-T. Hu. Homotopy Theory, Academic Press, New York, 1959.

[5] T. Karube, "On the local cross sections in locally compact groups," J. Math. Soc., Japan, 10 (1958) pp. 343-347.

[6] L. Keldys, "Example of a one-dimensional continuum with a zero-dimensional and interior mapping onto the square," Dokl. Akad. Nauk SSSR (NS) 97 (1954) pp. 201-204 (Russian).

[7] P. S. Mostert, "Fiber spaces with totally disconnected fibers," Duke Math. J., 21 (1954) pp. 67-73.

SOME QUESTIONS IN THE THEORY OF

NORMAL FIBER SPACES FOR TOPOLOGICAL MANIFOLDS

by

E. Fadell*

1. Introduction

1.1. Let M denote a topological n-manifold (separable metric, connected locally euclidean space) imbedded in a topological (n+k)-manifold S. Then, a tubular neighborhood of M is an open subset U of S such that U M and a map $p: U \to M$ such that the inclusion map $i: M \to U$ is a section for p and if $U_0 = U-M$, then (U, U_0, p, M) is a locally trivial fibered pair with fiber (R^k, R^k-0). It is clear that a necessary condition for the existence of a tubular neighborhood U of M is that M be a locally flat submanifold of S. (In the category of differentiable manifolds such tubular neighborhoods always exist.) Whether or not this condition is sufficient is still an open question, i.e., we do not know that locally flat submanifolds possess tubular neighborhoods. Such tubular neighborhoods would, of course, serve as the "normal bundles" in the topological category. If one relaxes the requirements of what these "normal bundles" should be, then it is possible to associate with locally flat imbeddings and immersions a gadget which has many of the properties desired of a "normal bundle." Such a gadget we refer to as a normal fiber space and we review the precise constructions in the succeeding sections and raise some questions. Details on how the theory works, paralleling the theory of characteristic classes in the differentiable category, may be found in [3] and [4].

2. The Normal Fiber Space of an Imbedding

Let $f: M \to S$, denote a locally flat imbedding of an n-manifold M in an (n+k)-manifold S. We identify M with $f(M) \subset S$. Let N_0 consist of all paths $\omega: I \to S$ in S such that $\omega(t) \in M$ iff $t = 0$,

* Work on this paper was supported by the National Science Foundation under Grant GP-3857.

$0 \leq t \leq 1$. Let N denote N_0 plus all the constant paths in M, and give N_0 and N the c-o topology (see subsets of S^I). Define $q: N \to M$ by $q(\omega) = \omega(0)$, and set $q_0 = q|N_0$. Furthermore, let $(F, F_0) = (q^{-1}(b_0), q_0^{-1}(b_0))$ for some base point $b_0 \leq M$. Then, $(\xi, \xi_0) = (N, N_0, q, M)$ has the following properties;

(1) (ξ, ξ_0) is a locally trivial fibered pair with fiber (F, F_0),

(2) $(F, F_0) \sim (R^k, R^k - 0)$,

(3) Let $G(S, M)$ denote the topological group of stable ([1]) homeo morphisms which leave M invariant and $G_0(S, M)$ the subgroup leaving b_0 fixed. Then (ξ, ξ_0) is a Steenrod bundle pair with structure group $G_0(S, M)$ and in fact the associated principal bundle is just

$$f: G(S, M) \to M$$

where $f(g) = g(b_0)$.

2.1. Question. Is there a euclidean bundle $(\eta, \eta_0) = (E, E_0, p, M)$ which is fiber homotopy equivalent to (ξ, ξ_0). Here (η, η_0) has fiber $(R^k, R^k - 0)$ and group $H_0(R^k)$, the group of homeomorphisms on R^k which leave the origin fixed.

2.2. Remarks. First of all, observe that the existence of a tubular neighborhood for M obviously implies an affirmative answer. On the other hand, it is unlikely that an affirmative answer to Question 2.1 will imply the existence of a tubular neighborhood. Also, it may be worth recalling the Nash tangent fiber space of a manifold. Given a manifold M, let T_0 denote those paths ω in M such that $\omega(t) = \omega(0)$, if $t = 0$, $0 \leq t \leq 1$, and let T denote T_0 plus all the constant paths in M. Define $\pi: T \to M$ by $\pi(\omega) = \omega(0)$, set $\pi_0 = \pi|T_0$, and let $(F', F_0') = (\pi^{-1}(b_0), \pi_0^{-1}(b_0))$. Then $(\tau, \tau_0) = (T, T_0, \tau, M)$ has the following properties.

(1) (τ, τ_0) is a locally trivial fibered pair with fiber (F', F_0'),

(2) $(F', F_0') \sim (R^n, R^n - 0)$, $n = \dim M$,

(3) Let $G(M)$ denote the group of stable homeomorphisms of M and $G_0(M)$ the subgroup leaving b_0 fixed. Then, (τ, τ_0) is a Steenrod bundle pair with structure group $G_0(M)$ and associated principle bundle $f: G(M) \to M$ where $f(g) = g(b_0)$. The question analogous to 2.1 with (τ, τ_0) replacing (ξ, ξ_0) is answered affirmatively by letting (η, η_0) be the "core" of the tangent microbundle of M (see Kister [4]).

2.3. Question. What favorable topological properties do the spaces (N, N_0), (F, F_0), (T, T_0), (F', F_0') possess? It is known that all these spaces are ULC (uniformly locally contractible [5]). This result is due to D. C. West. Are these spaces ANR (metric)? If not, we would have an answer to the following question raised by Dugundji.

2.4. Question. Is a metric ULC space ANR (metric)?

2.5. Question. Consider the tangent fiber space (τ, τ_0) in Remark 2.2. Then, a section for π_0: $T_0 \to M$ is the natural analogue of a 1-field on a differentiable manifold. What might be analogues of k-fields, $k \geq 2$?

3. The Normal Fiber Space of an Immersion

Let f: $M \to S$ (dim $M = n$, dim $S = n+k$) denote a locally flat immersion, i.e., for some open cover $\{V_\alpha\}$ of M, $f|V_\alpha$: $V_\alpha \to S$ is a locally flat imbedding. Because $f(M)$ may have self-intersections, the constructions the construction of a normal fiber space for φ presents more of a problem.

3.1. Definition. A non-trivial normal path is a pair $(b, \omega) \leq M \times S^I$ such that

a) $\omega(0) = \varphi(b)$,

b) $f^{-1}(\omega(t)) \cap V = \emptyset$ for some open set V (V depends on ω) containing b and all $t > 0$.

A trivial normal path is a pair (b, ω_b), where ω_b is the constant path at $f(b)$.

Let N_0 denote the set of all non-trivial normal paths and N the set of all normal paths, both trivial and non-trivial, and give these the c-o topology. Then, we have a natural map q: $N \to M$ by setting $q(b, \omega) = b$. Set $q_0 = q|N_0$ and let $(F, F_0) = (q^{-1}(b_0), q_0^{-1}(b_0))$ for some base point $b_0 \in M$. Then, the following is easy to establish.

3.2. LEMMA. $(\xi, \xi_0) = (N, N_0, q, M)$ is a locally trivial fibered pair with fiber (F, F_0).

Unfortunately, however, the following result is not difficult to prove.

3.3. LEMMA. F_0 has the homotopy type of $R^{n+k} - 0 (\sim S^{n+k-1})$.

If (ξ, ξ_0) is to serve as a "normal bundle" we should have $(F, F_0) \sim (R^k, R^k - 0)$, i.e., $F_0 \sim S^{k-1}$. This is accomplished (at least up to weak homotopy type) by altering the topology on N by enlarging the number of open sets. This new topology is obtained as follows.

Let V denote a neighborhood in M such that f: $\overline{V} \to S$ is an embedding and $\overline{V}$ is compact. Let $A(V)$ consist of all pairs $(b, \omega) \in$

N such that $b \in V$ and $f^{-1}(\omega(t)) \cap \overline{V} = \emptyset$ for all $t > 0$, together with all trivial normal paths (b, ω_b), $b \in V$. Let Φ denote the family of all such sets $A(V)$ and Ω the c-o topology for N.

3.4. Definition. The weak topology for N is given by taking the topology generated by $\Phi \cup \Omega$. N with this weak topology is designated N^*. N_0^* wil designate N_0 with the weak topology, $q^* = q\colon N^* \to M$, (F^*, F_0^*) is (F, F_0) with the weak topology.

One determines the weak homotopy type of the pair (F^*, F_0^*) as follows. Choose a sequence of "nice" neighborhoods V_1 V_2 $\cdots$ each containing b_0 such that for any set G open in M, $b_0 \in G$, then $V_j \subset G$ for some j. Let

$$\left(F^*(V_i), F_0^*(V_i)\right) = \left(F^* \cap A(V_i), F_0^* \cap A(V_i)\right) .$$

then,

$$\left(F^*(V_1), F_0^*(V_1)\right) \subset \cdots \subset \left(F^*(V_i), F_0^*(V_i)\right) \subset \cdots$$

gives (F^*, F_0^*) as an inductive limit of open subsets. All the inclusion map can be shown to be homotopy equivalences and $(F^*(V_1), F_0^*(V_1))$ can be shown to have the same homotopy type as $(R^k, R^k - 0)$. It then follows that (F^*, F_0^*) has the same weak homotopy type as $(R^k, R^k - 0)$ thus one obtains

3.5. PROPOSITION. $(\xi^*, \xi_0^*) = (N^*, N_0^*, q^*, M)$ is a locally trivial fibered pair with fiber (F^*, F_0^*) which is of the weak homotopy type of $(R^k, R^k - 0)$. (ξ^*, ξ_0^*) is called the normal fiber space of the locally flat immersion f.

3.6. Question. Is there a euclidean bundle $(\eta, \eta_0) = (E, E_0, p, M)$ which is fiber homotopy equivalent to (ξ, ξ_0)?

Obviously, a necessary condition for an affirmative answer to this question is that (F^*, F_0^*) have the same homotopy type as $(R^k, R^k - 0)$.

3.7. Question. Is $(F^*, F_0^*) \sim (R^k, R^k - 0)$?

3.8. What properties does F^* possess as a topological space? It is clearly Hausdorff. Is it ULC, ANR(Q) for some category Q ?

3.9. Remark. If, for example, one desires to apply Stasheff's classification theorem ([6]) to $q_0^*\colon N_0^* \to M$, one would need to know that $F_0^* \sim S^{k-1}$.

3.10. Does (ξ^*, ξ_0^*) have the structure of a Steenrod bundle? If so what is the structure group?

3.11. Even though (τ, τ_0) as defined in question 2.2 serves well as the tangent fiber space of a manifold, there is an alternative which works as a somewhat more natural "mate" to the normal fiber space of an immersion. Given a manifold M, let T_0' denote those paths ω in M such that for some $s > 0$ (depending on ω) $\omega(t) \neq \omega(0)$ for $0 < t \leq s$. Let T' denote T_0' plus the constant paths of M and give T' the c-o topology. As usual we have a map $\pi'\colon T' \to M$, where $\pi'(\omega) = \omega(0)$. Let $\pi_0' = \pi' | T_0'$ and (F', F_0') the fiber over $b_0 \in M$. Then $(\tau', t_0') = (T', T_0', \pi', M)$ is easily seen to be a locally trivial fibered pair with fiber (F', F_0'). However, just as in the case of the normal fiber space of an immersion F_0' has the wrong homotopy type.

3.12. LEMMA. F_0' is contractible.

3.13. COROLLARY. $\pi_0'\colon T_0' \to M$ always admits a section.

3.14. Remarks. Corollary 3.13 in the language of popular mathematics says that it is possible to cover the head completely with hair (well plastered down with that "greasy kid stuff") provided there exists at least on ingrown hair.

Thus, (τ', τ_0') is not acceptable as a tangent fiber space to M. We can, however, weaken the topology in T' in order to make things work. For fixed s, $0 < s \leq 1$, let

$$A(s) = \{\omega \in T_0' \colon \omega(0) \neq \omega(t), \quad 0 < t \leq s\} .$$

Then, we have

$$T_0 = A(1) \subset A(\tfrac{1}{2}) \subset \cdots \subset A(\tfrac{1}{n}) \subset \cdots$$

with $\cup_n A(1/n) = T_0'$. We introduce a new topology in T' so that each $A(1/n)$ is open as follows. Let Ψ denote the family of all sets $A(s)$, $0 < s \leq 1$, and Ω' the c-o topology for T'. Let Ω^* denote the topology generated by $\Omega' \cup \Psi$ and let T^* denote T' with topology Ω^*. Furthermore, let T_0^*, F^*, F_0^* denote, respectively, the sets T_0', F', F_0' with topologies induced by Ω^*, and $p^*\colon T \to M$ is just p'. Now, it is not difficult to prove the following.

3.15. PROPOSITION. F^* is contractible and F_0^* has the weak hanotopy type of S^{n-1}, where $n = \dim M$.

3.16. PROPOSITION. $(\tau^*, \tau_0^*) = (T^*, T_0^*, p^*, M)$ is a locally trivial fibered pair with fiber (F^*, F_0^*) which has the same weak homotopy type as $(R^n, R^n - 0)$, $n = \dim M$.

3.17. Definition. (τ^*, τ_0^*) is call the extended tangent fiber space of M.

3.18. Remark. (τ, τ_0) and (τ^*, τ_0^*) are weakly fiber homotopy equivalent, i.e., the inclusion map α: $(\tau, \tau_0) - (\tau^*, \tau_0^*)$ is a weak homotopy equivalence when restricted to fibers. If we know that $(F^*, F_0^*) \sim (R^n, R^n - 0)$, then we could assert that α is a fiber homotopy equivalence.

3.19. Question. Is $(F^*, F_0^*) \sim (R^n, R^n - 0)$?

3.20. Question. What topological properties are possessed by the spaces F^*, F_0^* ?

REFERENCES

[1] M. Brown and H. Gluck, "Stable structures on manifolds I," Ann. of Math., 79 (1964) pp. 1-17.

[2] E. Fadell, "Generalized normal bundles for locally flat imbeddings," Trans. Amer. Math. Soc., 114 (1965) pp. 488-513.

[3] E. Fadell, "Locally flat immersions and Whitney duality," Duke Math. J., 32 (1965) pp. 37-52.

[4] J. Kister, "Microbundles are fibre bundles," Amn. of Math., 80 (1964) pp. 190-199.

[5] J. P. Serre, "Homologie singulière des espaces fibrés," Ann. of Math., 54 (1951) pp. 425-505.

[6] J. Stasheff, "A classification theorem for fibre spaces," Topology 2 (1963) pp. 239-246.

UNIVERSITY OF WISCONSIN

Milton Keynes UK
Ingram Content Group UK Ltd.
UKHW022336260224
438516UK00001B/34

9 780691 080567